AF505774

COSMIC
WONDER

378-RUSS

COSMIC WONDER

Laurelle Russell-Atkinson

378-RUSS

Copyright ©2000 by Laurelle Russell-Atkinson.

ISBN #:   Softcover   0-7388-2862-9

All rights reserved. No part of this book may be reproduced or transmitted in any form or by any means, electronic or mechanical, including photocopying, recording, or by any information storage and retrieval system, without permission in writing from the copyright owner.

This book was printed in the United States of America.

To order additional copies of this book, contact:
Xlibris Corporation
1-888-7-XLIBRIS
www.Xlibris.com
Orders@Xlibris.com

Contents

ch.6 a cosmic code?

ch.7 a context for superluminal thinking

ch.8 cyclical links

ch.9 superluminal travel

Preface

The final suggestion proposed for superluminal power is an example only of what you can come up with using the book's theories. Which is to say that whether or not you can strum up superluminal propulsion from manipulating proton spin is questionable. And if indeed you can't do it that way, so what. I don't think that should invalidate the book's theories. Rather, I'm hoping these theories can open windows of thinking space for determining such possibilities. There could be many different ways to realise such power if at first we broaden our cosmological models and frameworks for reasoning. If we can rise above the present limitations of the known laws of physics we're on the way. Now that's what the book's all about.

Introduction

Tasmania 30th December, 1999

At last, the book is all but finished. I'm just going over the beginning and that should about wrap it up. Or is this just the beginning? I might need to take another look at chapter 1, and 2 . . . . And nothing is in print or published yet. One can only enjoy what's left of the writing before all that hard yakka starts. I hope you enjoy reading it and half your leisurely luck if you do. I've had computers blow up—jolly viruses in the works. Be blowed if I know how I'm going to print it. Well, so what. It's all par for the course, I guess. More importantly, have you ever wondered about distant galaxies? I wonder if we could live up there. Why can't we go there and what's stopping us? The speed of light, you think? Maybe. But not having an ocean liner and a mobile never stopped Columbus. I suppose uncovering the speed of light though is one step on the way. I mean until you know what you're up against how do you overcome it? The thing is we're not overcoming it. Having found out about it; we're letting it set in concrete around us. There it is. Man's all time discovery—his own prison walls. We're in a bubble below the speed of light. All physical phenomenon stops at that speed for us, and we're discovering nothing beyond it. That's how it's looking. So what if we step out a bit in our wasteland of a solar system. We don't get much beyond it, and there's bugger all in it. We're in Hubble's bubble as far as we can see. And who's coming here to say otherwise—no-one that I can tell anyway. The closest we seem to get to distant galaxies is a Star Wars movie, or a meaningless photograph of coloured lights. Maybe it's just as well, then. Nevertheless I'm writing a book about it.

Mind you, coming from a limited background in philosophy and political science, means the following is fraught with apologies—not to those disciplines, rather for my failings. You're dealing with a mathematical midget forthwith. I've only got a belief, not a degree, in all of those stars and galaxies up there. I believe they're all part of an infinite and eternal cosmos. It's that simple. And from that point I formed an idea for getting there. It's a theory for superluminal speed. I reckon it's the only way we're going to escape this bubble (superluminal speed that is, not necessarily my theory of it). Now simple as it basically is, when you get into it, other issues necessarily arise. And for an unqualified cosmologist—well let me just say that at least they don't phase me. But surface they surely do, like an iceberg, and I know what I'm grasping is small change to those aware of their greater depth. Without the maths and physics I'm explaining things to myself as well as you to get the theoretical points across. And then I sometimes have to explain them again to make certain. I hope I've got the main ones right. You see; to overcome the speed of light apparently means overcoming the greater quandaries in contemporary cosmology.

Now having said all that, I hope, hasn't put you off. What sort of book is this? It's different. It's not your standard science. I don't see it ever standing up in front of a class room of students. It's no note of authority. But do you really think we're going to overcome our laws of physics if we only ever stick with that? The ace up my sleeve is what I don't know offers plenty of free thinking space. And that's bigger than the Milky Way. Look, revering these laws hasn't got us past them. And who really understand them anyway? Who really knows how far they apply? Our powers of observation limit us immensely cosmologically. So if nothing else, the following bravado might highlight some limitations to these laws—possibly through an inadvertent grasp of objectivity. We may need to construct a different platform, for instance, for understanding the universe better, one that I've cottoned onto.

At university they told me we're realists living in the post-modern era. And so we are. It smacks of a time-space compression by way of yesterday's extension. All they do is go into the immediate realities of tomorrow. You don't get much beyond it, you just sort of tune in to the present more. Live for today. This is where understanding has no space for distant connections. Nor does it have the time for universal necessities. You can see our nature becoming increasingly subjective. Well, I can. I get it on the television every night in some form or other. It generally amounts to me-now-very-fast, and no please about it. We've probably consolidated our choices into the egocentric range. Our values are looking transitory and hedonistic. Maybe that's the power of full on free market forces and capitalism. Maybe that's the speed c limit and the Hubble bubble taking effect. Whatever. I'm thinking our cosmological concepts are partly why. They could be projecting a limited dimension. And that could be blocking evolution.

Cosmology, a great exemplar of the scientific age, looks like a field clogged up with barriers and dead ends. All you seem to get are paradoxes, uncertainty, and impossible distances with it. You've got no pre big bang info. You've got the 2 main strands of thinking in discord. Look, we're probably judging the scope of an ocean using the power of a cross-eyed 2 legged ant. That's the speed c barrier for you. Indeed, cosmology for all it's worth as we're getting it, probably just encourages cynicism, end-of-science thinking, and apathy. I'm hoping to rise above all that. By delving into our scientific edifices, the laws of physics, it's easy to recognise our potential. At the same time, it becomes increasingly clear how much farther there is to go for us. But you might as well forget it going by the way things stand today. There's a lot of jaded and cynical minds out there. Besieging us are the remnants of broken dreams. We're feeding off them. Tune in, turn up and strum up the reality of another one, rather than realise an ideal. Just listen to the music. So where do you get the power to go the distance?

Don't look at philosophy. And I'm not getting into the ebbing currents of religion. All you're getting philosophically is post-modernism; post-structuralism; de-constructionism; existentialism, realism—4 syllable isms that generally mean nothing to anybody. Maybe realism registers. Anyway, these things tie in with democracy, capitalism—the politics of the day. You'd have to be an Ishmaelite to identify with any idealism now, and maybe that's only syntactically. You don't see them anymore. Who knows, maybe their realities swamped their ideals as well. The thing is; idealism could well be just another powerless esoteric term. And with all our hi-tech wizardry we could be in a post-modern rut of our own peculiar real world. It's real and fast. It's happening all around. Your broken ideals of yesterday render tomorrow's tomorrow out of focus. No one cares. But what if you did? What if you got an idea for inter-galactic travel, for instance? What if this idea stayed in your mind for sometime, not to be distracted from by every other distraction going? On what basis could you start?

Where for example, is the basis for objective reasoning in existentialism? Maybe I misunderstand it. Where's your platform of morality? Where's the divine inspiration of eternity and infinity? They shut the window, and now you're looking at a pane of glass reflecting your own image, instead. We've got everything up for observation and experiment. It all goes through our microscopic speed c range. Now how can everything do that? There's more in the cosmos than we'll ever see. And that's your basis—the scientific below speed c method. You might as well forget your forest and just focus on the tree then. How are you going to transcend your tree if you can't see where you're going? The cosmic forests I'm talking about go beyond the observation and experiment of any speed c limit.

A concept of eternal infinity could be a major start for galactic contemplation, or is that vice versa? Either way, without considering eternity and infinity as a guide for the reasoning process, our understanding of evolutionary processes could be lacking. But we seem to decide all possibilities from the subjective position. There seems to be no objective platform from which to form a decision, and no need to work out eternal infinity. Now with cosmology, I think a concept of infinity and eternity is vital. How else are you going to get a 'Theory of Everything'? Clearly the conceptual approach is the only way to go. Who's going to *observe* it first? Who's going to catch a glimpse of eternity and infinity and then hit the drawing board, Moses Mark II? Where is he? Then, again I suppose you *can* do it. If you look at the galaxies long enough and believe in their depth I suppose there's your glimpse to get you going. Looking at them often enough might encourage some conceptual thinking in their direction. They develop a sense of wonder. Well, they do for me. Laurelle in Wonderland. This however, is the wonder of reality. Up there is a reality sufficiently rich for ideas. There's your platform. But the point I'm trying to make is that your concept of infinity and eternity would put some philosophical impetus into the rationale.

And here again is the point of the book, reaching for the stars and the way to go. The laws of physics enforcing the luminal limit are both proven and revered. So you're up for untold ridicule and scorn trying to get around them. Be that as it may, these laws could apply to stages of velocity, defining specific transformation processes. How far a law is relevant could depend on how far we want to take it. General relativity doesn't explain what happened before the big bang, and neither does quantum theory. Do we leave it there then? Do we stay within the reality of these laws? Do we stay in our 'known' cosmos subject to their limits, which seems to amount to a closed system, end of story. On their basis, we could remain in this solar system with nothing beyond our own intelligence for reflection. And what goes on behind the Magellanic Clouds is none of our business.

Our sun is less than half way through its life. Yet here we are as primitive as ever; still fighting and carrying on. The mentality doesn't change much. Indeed, our evolutionary progress within the universe seems limited by our earthly mismanagement, rather than the solar life span. Maybe our planet will become another wilderness in this solar system, still young yet burnt out before its time, like a lot of its residents. And all because we have no idea. It's up to us, as the long history of science shows. Future reality may depend largely on our ability to imagine and create and fire up the scientific method beyond speed c's snail pace.

So you've probably got to transcend the scientific method to get the ball rolling. Look, even I know that sometimes recognising limits in knowledge comes from beyond the confines of the discipline in question. It's academic. Even so, without input from a variety of disciplines how can we know an 'objective truth' or apply it? Physics, for instance, might only apply the findings of observation and experiment. This then forms the basis for cosmology—the study of a universe that is greater than our observational abilities? Shouldn't it also require some degree of extrapolation and speculation based on theory? And I don't doubt that it does already. My point however is: the focus is way too narrow. And physics alone may be in no position to broaden it. The underlying paradigm could be imposing too many finite boundaries, leaving cosmology stuck in gear at light speed, or the ones below it, or even in reverse. What would happen if we apply a different concept to the limits of knowledge in cosmology? This could place the known laws in a different context. Maybe we could resolve the disparities between general relativity and quantum theory, for starters. Perhaps we could figure out a continuity to universal processes.

Take the multi-disciplinary approach. That's me. Using disciplines outside physics, I can apply an idea of infinity and eternity to the observational evidence, which I try to do through out the text. Using maths and physics however, I probably wouldn't come up with the same things. I mean an equation isn't a question. Besides, I'd come up with nothing, being mathematically deficient. And even if I could, who's reading all of those numerical expressions?

My bent is philosophy and international political science, both of which place the emphasis on debate. The realism of one, however, can contrast with the logically structured ideals of the other. Philosophy allows you to delve into the greater unknown—the one that physics seems to withhold. Nor does it confine the debate to the scientific method. You can be more creative with it. At the same time, I've found the objective arena of international political analysis keeps the focus on the reality of the day. And because they're generally pretty grim, you can find yourself philosophizing their resolution. In other words, I have learnt that to overcome realities sometimes you have to start from a position transcending them. Now all of this is what I'm applying to cosmology.

Still, let's hope Cosmic Wonder isn't the blind presumption of a parvenu. I can only say that the deeper one gets into cosmology the greater respect I have for every jolly superior in this fantastic field. That's a lot of respect. The physicists impress me the most though. Too bad I can't join forces. Would Bernhard Riemann collude with Bugs Bunny over carrot juice? Not to worry. No-one's yet applied a concept of eternity and infinity to the universe that can explain beginnings and endings in a regular pattern. Number's still haven't explained it, despite being heralded as the universal language. Besides, no offense Riemann, but the smartest mathematician amongst us could have the algebraic skills of an echidna compared to what's up there. For all my ignorance, it seems we're still in linear land mathematically, anyway. How do we know the data before us doesn't extrapolate systematically through eternal patterns of cycles? I reckon we're in a cosmos full of nonlinear systems. The solar system looks nonlinear. So do the galaxies. So do atoms. Strange then that the whole business should start out from point zero.

ch.1

questioning : the laws of physics
            : the speed of light limitation
            : the spacetime concept

alpha and omega

Today, how we reach understanding may have little to do with a belief system. The emotions involved in thought processes that predicate understanding with belief seem lost and irrelevant in the real world. What happens after we die, is perhaps inconsequential to an increasingly subjective nature. Anyway, history shows that we tend to think through observation. That, along with our subjective nature, could be partly why we don't understand the universe as infinite or timeless.

Our concepts claim universal expansion began with time. But has it proved an adequate way to understand the cosmos? We could be invalidating any position for determining eternal cyclical phenomena. For example, as the 4th coordinate in Einstein's equations, time conceptualises as having a beginning. The universe is then apparently, understood to evolve from the zero radius point of time. Which then renders space unlimited? Baloney. Space here begins with time. And time's got the beginning. Where's your endless past with space? Nor do you get all the dimensions that way. You couldn't reason transformation processes through a complete context of continuity either. At some stage things didn't happen. They just *started*. Where's your ongoing pattern of events then? Look, your spacetime concept could be expressing a teleological interpretation of events within an infinite mathematical world. It shows a final cause within infinity. You get the infinite part but not the eternal part. I reckon it's only describing part of a process. It only shows stuff emerging from nothing to full on everything. It ties in with a subjective position. So where's your order and systematic turn of events through the past into the future?

What would happen though, if time became inviolably linked though motion to infinity, then applied to space? Wouldn't this shed a different light on the dimensions of the universe? You might reach into eternity this way. Think about the dimensions of the biggest possible conceptual framework. You only get time where there is motion, and you only have space where there is time and motion. None of them have a beginning, or an ending. And none of them exist without the other two. So time never stands still. Where's your quest for determining the ultimate state and creation of matter then? There mightn't be any. We could learn how nature works through cycles and systems instead.

We could be discerning only part of a process through gravitational relativity. Clearly there is an order to nature through expansion. But we can't say for sure that this expansion covers the entire universe. It only covers what we see according to the general relativity of gravity. We can only get the spacetime and motion dimensions below speed c, according to gravity. We know that gravity isn't the only force, however. And that at some stage all these forces compressed into a micro dot. Indeed, it looks like they all emerged from a contraction of spacetime and motion. Who's to say the general relativity of another force didn't prevail then? Say one of the nuclear forces. Motion might've figured more prominently within space and time then. You could've had loads of motion in very little space and time. You could've had a general level black hole of overwhelming gravitational pull translating into the general relativity of spherical spinning strong nuclear force. You see; what we're looking at today could be the expansion *stage* of a cycle of spacetime and motion, over which the gravitational force prevails. One where space and time overshadow motion, amounting to a gravitational force below speed c. But the cycle—to come full circle—would involve three other stages, that of the other three forces. If you want them to add up, they should each prevail at some stage or other. Why just gravity? And these cycles of force predominance might be occurring not just before and after each other. They could also be happening next to each other. With the bigger picture you've got more room to move. You can locate the dimensions where causal processes manifest systematically. Some inter-galactic systems, for instance, might be contracting, strumming up a general level force. Maybe we're going out while they're coming in, like a load of figure eights.

Is eternity and infinity really so far fetched? Let me put it this way; if I can work something out through it, anyone can. And if thus far, it's looking dodgy, then that's only because I'm not doing lucid justice to nature's inherent simplicity. The basic idea involves; drawing a symmetrical connection between the eternity of time and the infinity of space, with the presence of motion. Your symmetrical differentiation of spacetime and motion realises four forces. They contextualise through the systems of cycles. There's our concept in a nut shell. So symmetry's an issue in itself—well, I'm getting there. For the theoretical time being let's just say the balance symmetry affords creates systems within the cycles. The point here is; using a concept of infinity and eternity enables you to decipher systematic processes of the cosmos. And you don't need to look hard to find them, nature offers endless examples. We've got systems and cycles on a variety of scales. There's atomic ones, stellar ones, galactic systems. Why should it be inconceivable that the same applies on the inter-galactic scale? Clearly there's some general level. Just broaden your mind and contextualise it beyond our observational boundary. I'm suggesting they too form a systematic process, a.k.a. cycles. They all should reflect an evolutionary pattern of life.

There's probably some symmetrical pattern of eternal events that our knowledge to date disregards. Our present laws of physics, subjectivity and observation wouldn't go the distance. And should you give it a whirl, infinity and eternity probably conjures up images of vast timeless nothingness, or chaos. Neither of these scenarios rings true. The deeper you get into it, the clearer it becomes. The fabric of reality—to borrow a term—of eternal spacetime and motion, shows a symmetrical pattern. Everything in the universe could necessarily interconnect. I'm thinking there is a cosmic design of eternal processes of matter transformation. Fine. How much of that though, can anyone prove? And this is what makes drawing up a theory on an infinite and eternal cosmos so enigmatic. It's the whole jolly point of the book. You see, it seems to link in with the speed of light. If we can see beyond it, for starters, into more distant realms we might be able to prove things through hard evidence as well as extrapolation. We could get the grand-scale cyclical pattern of events that accord with an infinite and eternal cosmos.

Do we get booked for exceeding the speed of light?

Now before getting into this, I've got more questions to ask. I don't think you can make any superluminal sense without first setting the scene. Look, are we really in a position, in our evolution, to say nothing in the entire universe travels beyond light speed? Or, that it all began with a big bang? The speed of light limits our powers of observation and experiment to a micro-cosmic smidgen of what the cosmos is all about. Who would disagree? Even Hubble's brilliance doesn't get us beyond the boundaries of a bubble. We don't know if the universe is consistent. Nor do we understand cosmic evolutionary patterns. Why do supernovae explosions appear infrequent to us? Why does the matter left in their remnant cores have an apparent gravitational pull so great that it defies the speed c limit? Why are we still in this solar system? We haven't even realised an harmonious and efficient existence on this planet. No jolly wonder. There could be many reasons, apart from our laws of physics, why we remain here within the speed of light.

Through our time and motion laws, space is not unlimited for us. So what if we can travel through a myriad of times and distances, in many forms below speed c. Most of them only churn out pollution and traffic jams. Beyond speed c, time and motion have no meaning. Therefore neither can space. This is where time stands still for us, and has an ending, which is inconceivable if the universe is in constant motion. Perhaps it's just easier to think through finite concepts, then to comprehend the eternity of time and motion, or any belief system.

Nevertheless, how we use time and motion in our working frameworks, should produce a specific result from a set of possibilities. And the one we've got now might have effectively closed off all avenues for understanding motion beyond the speed of light. Take Einstein, for example. His dream was to understand all the forces of nature as different aspects of a single fundamental force. Yet he still restricted the causal rules in space and time, to gravitational relativity. That forbids any physical influence from propagating faster than the speed of light. That means forming no relations beyond that level. And it makes the speed c a barrier with time part of its structure. One force, namely gravity seems to co-ordinate with one, electromagnetism, and over rule the other two. The thing is; gravity is only one force. It just happens to be the predominating one in our expanding vicinity. There are three other ones though. And if the cosmos functions systematically, at some stage they should show up more effectively. Under the predominance of gravity, and it's structural guidelines, we couldn't detect any superluminal dimensions. But imagine Einstein's dream. If we untangled the differentiation of one force into four we might find ways around this. First off, we could try *contextualising* gravity's limitations of spacetime and motion within a bigger framework of spacetime and motion. And it's not that hard because Einstein's done all the ground work.

Both mass and energy, as measures of the same physical quantity, are also subject to a common limit; speed c. Their conversion factor though, goes up to speed c *squared*. Nevertheless, that doesn't appear to illuminate any unifying purpose between the forces of nature. But I think it should. Mass and energy are still measures of spacetime and motion. Spacetime and motion differentiates into four forces. One is subject to the other. In other words, forces relate to mass and energy transformation processes. And those processes relate to forces. Yet we have limited all of these factors to the effects of one force only. That's one whose effects we can only discern through a limited spacetime framework anyway. We're restricting the conversion equation to this framework, whatever the force.

And you might say, well it works. But it doesn't completely work. There's big disparities between nuclear forces and the general level of gravitational relativity. It's like somehow we've managed to find out bits and pieces of cosmic mechanics, but only enough to get around the solar system. Heaven forbid that we figure out the bits in-between and get out there. Heaven probably does.

. *cosmic consruction and at the frisbee level*

The principle of relativity apparently defines through physics a way to understand the *process* of change in electromagnetic phenomena. I read about it. That's how it seems to go into the atomic systems. But it doesn't add up inside the atom. They don't mix. You've got big disparities between quantum theory and general relativity. Why should grand scale gravity reach *inside* the atom anyway? I don't reckon it does. You're only presuming a modified dimension of one form of mass/energy, say beyond the atom, or atomic mass plus, applies to another, say sub-atomic. Greater atomic relativity could be restricting our knowledge of inner atom relativity. And applying it across the board probably conceals intricate continuities of both. Continuities that possibly relate to nuclear forces. We don't seem to get the entire transformation process between mass and energy. We don't get the sub-atomic and atomic connections clearly. Our knowledge of position and momentum of sub-atomic stuff for instance, falls short. We get one or the other. The sub-atomic set-up of electromagnetic phenomena, however, could also be subject to a system within itself. That means four-force activity. Perhaps electron orbitals have nothing to do with macro gravity. They could primarily attune to some micro system, ergo the atom. What I'm saying is that the atom could happen through its own systematic process involving four forces. For me, the electron orbital signifies sub-atomic gravity. And I can see you scratching your head here. Look, we're only theorising. Still, who's to say that's not the case? The gravitational relativity of a grander system might only affect the external mechanics of electromagnetism. That's what escapes the atom.

Our applied framework would only measure what fits into it. Yet the transformative nature of matter and energy should concern four forces. And the nuclear forces transcend our conceptual limitations at certain levels. They don't always fit into the generally relative framework. O.K., so we get the quantum theory for nuclear activity. But that means two different frameworks. And the new nuclear one still doesn't recognise any superluminal dimensions. Nor does it align with the gravitational framework. General level gravity won't fit in with the nuclear dimensions and vice versa. So how do you know interactions between sub-atomic phenomena don't occur above speed c.? We could be dealing with energy levels beyond both frameworks. And this could be one reason why the two theories/models don't align.

The point is; your process of continual change in electromagnetism, the energy and motion of it, doesn't describe continuity between nuclear activity and gravity anyway. And it looks like the speed c barrier is the obstruction. So I'm wondering there are different motion limits we can accord to different forces—as symmetrical subtelties of the one, most of which are beyond detection through the greater gravitational one. You might only pick up the range inside the system. Each force might involve a discriminating combination of energy and mass that pertains to a definitive combination of spacetime *and motion*. Electromagnetism happens at speed c. Gravity happens below it. The short range intensity of the nuclear ones might go above it. You see; although we only know about the short range intensity of nuclear ones inside the atomic shell, it could just be the shell per se that stabilises below speed c according to the general relativity of gravity. And your radioactive release could always be at speed c.

We don't grasp the nuclear intensity of say a quasar, supernovae or black hole. We don't get the four force system on the general level. For us, at high energy chaos replaces the laws of physics. We know that nuclear binding forces are much stronger than both electromagnetism and gravity. But how particles gravitationally bond together at high speeds is a mystery. Is this because we apply the speed c limit right across the board? Well, through the board to be more precise. Just look at the wave/particle dual nature of sub-atomic stuff and the uncertainty principle. That happens at speed c. Beyond speed c we might find formations that render the wave/particle set-up the simplicity of play dough. Try working out a spiral along the linear lines of the spacetime continuum, for instance. The point being that the full transformative nature of matter and energy may not necessarily stop at the wave formation of speed c. We just get it that way. We can't see otherwise. We only detect the gravitational effects on the grand scale. And we only find the nuclear ones on the smallest level.

Evidently the cosmos comprises of systems within systems. Atomic systems exist inside stellar systems. Then they go into galactic ones.

I'm suggesting the general relativity of gravity derives through the biggest. It's probably an expression of a particularly grand scale system. On that basis, it might only affect the external mechanics of internal systems. It could also obfuscate things. We mightn't detect the internal mechanics of more intricate ones through it. And for that matter, macro superluminal activities far out there deep in space are beyond us. These things could form the full transformation cycle of the general relativity system. Nonetheless, because of its speed barrier, and how we apply it, we wouldn't get beyond gravity's role in the bigger picture. Around the big-bang time this greater system could be under the general relativity of the nuclear forces. I've said that before. And it comes up again further into the text. It's fascinating stuff. For now, general relativity shows a unification between matter and energy. But it also shows a demarcation between the systems of nature, as in quantum systems with gravity. The general relativity of gravity no doubt connects all of these systems. But there still remain loads of inexplicable disparities.

Our entire understanding of waves, particles or anything else seems subject to the speed c of general relativity. That it is important is not the issue. Rather it's the arbitrary nature within a fixed context that I find questionable. Neither the framework nor the speed limit explains everything. What do we get at the borders of our concepts and dimensions? Nothing? We get high energy chaos. Or we fall into a black hole conundrum. Both of which still express motion in space and time. What does that mean for the spacetime continuum?

Clearly there are boundaries in nature. The ones we're uncovering through our laws of physics though look more like barricades. They project them as unfathomable. Our knowledge of the sub-atomic particle is uncertain and remains so with the uncertainty principle. What's with the wave/particle dual nature? The atomic system is in values of uncertainty. What's with 2 different working frameworks? What do these boundaries mean? So many questions. So few answers. Could they represent decisive levels within a bigger process? One that covers superluminal dimensions of spacetime and motion. One that involves an infinite and eternal cosmic nature. Let's put these boundaries into a broader context. Let's jump over the spacetime hurdle, for starters.

Interestingly, the conservation of energy and matter seems to reflect knowledge of continuity. As energy and mass changes from one form to another it has no definitive beginning and could be subject to all manner of endless motion. You don't *create* the stuff initially. Only God, the great eternal and infinite unknown, can do that. Anyway, it's always transforming in and out of each other. The continuity of mass and energy should embrace a continuity *with motion in space and time*. I don't see that the whole business ceases at light speed. Nor do I take 'time' as the primal cause of everything. There's a full circle process here that we're not picking up. And that's despite discerning a conservation to everything. Where's the conservation coming from? I think we should place it, and what it represents, within the greater context. Right says Fred, climbing up the ladder . . . So how do you contextualise perpetual spacetime and motion? If we remove the ceiling. . . .

That's one of those issues that develops along the way. There's an entire chapter later on about paradigms. There's also stuff on cycles, and more to the point now, in the next section—a lot on a spacetime *and motion* concept. It affords a bigger context. To this end, I'm back in our micro world. Why doesn't it accord with a general level, or make sense of continuity in processes?

. *waves and particles*

There's a value to the wave or particle nature of the sub-atomic particle. And whatever it is should tie in with an invariant constant of transformation processes. I worked that out some years ago. But it didn't ring any bells. Nor did it impress anyone, other than yours truly. What are these invariant constants, anyway? I'm suggesting those barricading borders form some of them. Basically it's symmetry in motion. All we know for sure though, is the links between particles and waves in quantum systems are uncertain. They're an increasingly complex and disordered mess to physicists. I wouldn't even attempt to grasp them according to our laws of physics. I think their overlap in terms of state and motion transcend these laws anyway. A beam of particles has wave-like properties. And although we can determine waves as a succession of particles in oscillation, we don't always find waves thus. We don't always find the particles to oscillate. When we do, their proliferation seems to produce increased complexity within one context, instead of any clarity and links to a higher energy order.

Photons, for instance, seem to describe these limits. They've got qualities that fit the physical nature of particles. Say, of energy, momentum and wavelength. Yet they're electromagnetic by definition and only get around at the speed of light. For all we don't know they could be linking energy from one context to another. I mean they form inside the atomic system and transpire out of it. And you're getting your general gravity outside the atom, for sure. But inside it, they could mark a transformation stage between energy and mass according to the atomic four force system. Sub-atomically, these things should link in with a nuclear combination of spacetime and motion as well as a sub-atomic gravitational one. Sub-atomically, your photon should have orbital affiliations as well as proton spin relevance. Look, any sub-atomic interaction should produce a different transformation of mass and energy. You might get electrons at the orbital end and you might get neutrons at the nuclear twisting of spacetime and motion. However, as we judge it under the general relativity of gravitational attraction, we'd get none of that. We reason atomic business through the stabilisation of blanket gravity. We don't see any superluminal potential inside the atomic shell. We don't see the motion of speed c as a symmetrical constant, systematically applicable inside and out. This motion is something that relates to the space and time of a wave *formation*. Your wave can form inside an atom and stay that way outside it—at speed c. It's a value of the nature of the photon.

Inside the atom the electron links in with orbital corridors of spacetime and motion. Here's a version of gravity happening *within* the atom. And inside the atom you're not getting any overwhelming general relativity. The other forces figure pretty well along side it. I'm calling it the atomic system of forces. Now when the electron moves into the orbital corridor of spacetime and motion, it would slow down. You're going into the *gravitational combination of spacetime and motion*. It's a constant combination inside or outside the atom. I mean your force is your force, and gravity is always below speed c. I'm just distinguishing between the systematic processes of them. Inside the atom you might find a mass position for your electron in an orbital corridor of spacetime and motion. That's mass, pertaining to a position afforded by the atomic version of gravity. That's a fixed combination of spacetime and motion. Outside it you'd find an atomic position—the system's position for mass according to a greater version of gravity.

The thing is, inside the atom the system of forces seem to fire away together. You're not getting a generally relative prevalence of any one over the others like you do on a bigger scale. The atomic gravitational orbital is probably subject and susceptible to other force interactions all the time. Pin it down and you lose the momentum. But take your electron out of an orbital and it should speed up. It can *spiral* through the dimensions of weak nuclear force spacetime and motion. It's going into a different formation. Energy, on that basis, would derive through less space and time and more motion. Shrinking down your space and time is rendering motion and its set-up more nonlinear. We're talking superluminal now. Once you start to spin beyond the wave frequency/formation, you're above speed c. So forget about position. You'd be lucky to accord it momentum. Perhaps it even spirals in and out of orbital corridors. There's loads of possibilities here. Anyway, electron spiral potential could reflect a photon intake of energy. I mean the electron's energy isn't just about mass and motion in this situation. It's about four nonlinear interactions within the atomic shell. It's about a nonlinear fusion of mass and motion. And when the photon exits the atom, it could retain some impetus and imprint of the superluminal format. Your particle reflex of photon activity could relate to its sub-atomic spiralised energy levels.

So you may be thinking you're joking by now. Photons, phooey. Electrons, forget it. Well, you can't write a book like this, just about writing a book like this. I've got to go into it. No superluminal wonderland is going to transpire out of my latest batch of home brew. How I wish it would, but try as I might, the only thing happening there is better beer. I could wax lyrical on it, but the fact remains photons transfer energy more than grog. And they do it via orbital electrons. I'm no authority, mind. I know none of these things for sure, even though it's probably common knowledge. Either way, if they're boring you, or what's more likely the case, if my desecration of them is sending you to sleep, well, there's other stuff further along. For superluminal travel however, these things look important. Post conclusion, I can tell you that it's all within the nucleus. And I think the technical ins and outs might get easier then if you're familiar with my weird understanding of these things. If not, then all I can say is, equations or beer anyone?

Photons can strike electrons producing photon emissions *without time delays*. That's in no time that we can perceive. And they change the energy level of the orbital electron. I think it revs it up by transferring oscillatory motion. Or is that spin? Now that's the stuff of the nuclear forces. A decrease of photon wavelength could transpire into the spinning speed of the photo-electron, placing it in a new energy-speed level. It could become subject to spiralised mass and energy, theoretically. One that pertains to the spacetime *and motion* combination of another force, say a nuclear one. And the time of that combination could be as imperceptible to us as its motion.

Either way, it seems that the wave-particle dual potential is always within the electron, and thus the photon. What then remains invariant, after the transformation? What is it about the electron that doesn't change? And ditto the photon. This invariant value could be necessary for nature to work within infinite transformation processes. It could amount to a particular combination of spacetime *and motion*. It could amount to a force *formation*. And as there are four forces there could be four formations here. That's four quanta of spacetime and motion in a nut shell. Their subtle interplay would produce different energy and mass transformations. Moreover, their subtle interplay could happen within its own little system. Or it could happen within its own enormous one. However, because we limit our laws of physics temporally, we don't evaluate processes thus. We only get the spacetime links of macro gravitational relativity. And because they're also subject to that motion barrier, we don't work out higher energy values alongside mass, or vice versa. We wouldn't get any particle position within a wave at speed c. Nor would we arrive at any mass association within a different energy formation beyond speed c, say of a spiral.

. *expanding the temporal dimension*

Measurements of universal phenomena clearly transcend our laws of physics. And these laws don't account for an infinite or eternal cosmos. The measurements we're getting through them show quandaries and uncertainty. Already we have the uncertainty principle. Maybe we'll get the quandary one next. We know there's similarities between the sub-atomic field and the planetary one. These could express symmetrically: a state or system that has a significant quantity that remains invariant after a transformation. (I read the dictionary.) Well, this symmetrical structure, evident through the causal process in both the micro and the macro systems, may be infinite in nature.

Rationalising the symmetry of systems, quantum or otherwise, through our laws of physics possibly won't show a transformative value. For one reason, symmetry is going though a framework that has a definitive beginning and maximum motion. So anything symmetrical between particles, energy, and forces could translate as; an inverse relationship between structures. (Refer; the other meaning in the dictionary.) That's how we might understand symmetry in the cosmos. Because of these limits, physicists—and theorists, could apply symmetry as a reversal of the time order. As such, they mightn't recognise other dimensions of spacetime and motion. They mightn't get the continuity flowing within a system. They wouldn't get the systematic interaction of four forces. They wouldn't draw links between the macro and the micro in terms of four force interaction within each. Nor would they see the impact one has on the other through the greater cyclical processes. Impacts, that only the continuity of an infinite and eternal framework can reflect.

Without the framework to carry forward forms of measurement, theorists could apply a reversal of the time order. Instead they could come up with reverse processes. Or they could arrive at chaos. They mightn't perceive the invariant structures within the eternal transformation processes of cosmic phenomena. Stuff that applies right across the entire cosmic spectrum.

How could every jolly object have an exact opposite? Opposite Laurelles? Opposite Sooties? If you're going to get the small time you probably get the big time as well here. With eternal and infinite spacetime and motion though, an object doesn't need to back track. It's always transpiring forward. It might have more continuity within transformation processes, rather than transgress into it's inverse replica. Heroclitus would agree. He never put his foot in the river the same way twice. His river forever flowed forward, and even if it didn't his foot wouldn't have gone in the same way. The thing is; the invariance we perceive could mark decisive transition stages within ongoing transformation processes. If we see them on the small scale as well as the large scale, they could reflect different systems of force interactions. Even our anti-particles could be different sides of the same coin. They should still denote continuity according to a system. Our debris of atomic bombardments that we're analysing might amount to parts of spin—parts of the internal structure of nuclear mass and energy. This could be the stuff from intense superluminal spinning sets of nuclear spacetime and motion. We could be picking up recoil segments of superluminal spin.

We seem to overlook the invariant side to symmetry using the linear scale of present frameworks. I reckon symmetry would show the transformational links, rather than isolated negative identities. It all comes back to our framework for understanding the cosmos. And our concept of spacetime sets some of these boundaries. Our concept of space curvature applies time as a starting point to endless space. It thereby determines the geographical scale. This linear framework can't show time continually happening with space and motion.

As such, the *system* of four forces interacting might go beyond our framework. That means you wouldn't get the systematic effects of symmetry. Your time's got to go full circle through space and motion into other set-ups for the systematic approach—for the symmetrical transitions. Indeed, transcending linear mathematics could be symmetrical estimates. So says the mathematical midget. I mean these estimates should relate to links of transformation processes. But using our linear framework, transformation values wouldn't contextualise that way. We wouldn't decipher the stages and cycles that systems can realise. We'd be looking at a barricade rather than a balancing act. And it's no use looking for Einstein. The definitive basis for time still underpins his equations for reasoning spacetime.

But maybe that's just giving us some of the picture. You're only getting your general gravitational curvature. Unfortunately for theory here, Einstein's brilliance can calculate his curvature. I'm giving you a bigger picture, but I couldn't calculate my way out of a bus stop. And who's got the intelligence of Einstein today to go with the theoretical nonlinearity I'm on about? Deep Blue the computer? Forget it. Deep Blue's probably too busy working on war games theories. Something's got to think about how Nato maintains New York if a nuclear Slobodan pushes the button. You might as well go back to the beginning, when space began with time.

Given time, nothing became something, curving in the process. With infinity and eternity there is always something. There could even be other forms of spacetime *and motion,* that connect through different processes. That all encompassing beginning that Einstein's looking at, with below speed c motion, could just be the beginning of gravitational general relativity. He could be calculating the curvature of a new cycle. One that has connections outside his time frame. You might find a nonlinear set up deeper within it. You might find a place where spacetime and motion have a tighter curvature, analogous to a spherical ball of nuclear dimensions. And there could be a wave one and a spiral one in between, all amounting to a massive system of macro force formations. Indeed, these formations could be the greatest symmetrical differentiation of spacetime and motion. Our expansion phase of predominating gravity could be one stage of a cosmic balancing act.

Too incredible? Well, how did you handle black holes when they first came out? Full of incredulity, I bet. When you think about it, everything's incredible at some stage. The whole cosmos is. Superluminal travel maybe no more ridiculous than Doctor Spock's ears, given time. Some stuff proves to be true and the rest gets overtaken by reality. That is, until something else arises for consideration. And how are we going to get there without coming up with something else? There's plenty of evidence before us, anyway.

Galaxies and quasars reflect bigger systems. Why are their formations so mysterious? We mightn't be able to reach into their dimensions, far flung or otherwise, to see their cyclical relevance. Maybe a quasar kick-starts a galaxy. Maybe they involve the non-linear dimensions preceding a gravitational curvature of spacetime and motion.

With our present working concepts, how can we understand creative forces? How do we grasp the ongoing and infinite nature of the cosmos? We may be defining the universe as we see it and measure it, according to a speed limit and time frame that neither reveals its entire nature nor our continued evolution. Both theory and experiment demand that nothing goes beyond light speed. It's a fundamental of modern physics. We don't detect anything beyond it, and there's a limit to the observable universe for us because of it. Using our knowledge of light and motion with the present laws of physics, we couldn't observe, visit, or adequately understand our closest neighbouring solar system. And good luck for them there. We're still in the grip of Darth Vader.

ch.2

the concept of spacetime and motion

. *preamble*

Theory really starts happening now. First up is the concept for spacetime and motion. That should set the scene for thinking beyond speed c. I mean the concept hopefully structures some thinking space. And that's where the eternal dimension comes in—from the inherent motion in spacetime. You can get a lot out of that. Already four forces, as manifestations of motion and spacetime, are taking on specific formations. They transcend the spacetime curvature of general relativity. We're not just contending with orbital formations and expansion phases anymore. Nor are we catching tangible spectres through the cross winds of a wave formation. I'm taking the linear plane into the nuclear intensity of spherical spin and spirals with this. We're going nonlinear. That means you get four forces in the one format. Linearity, after all, still fits inside the nonlinear dimension

You see; incorporating motion within the spacetime concept provides this theoretical nonlinear dimension. Using it, we can cover greater areas. Theoretically the concept allows us to look past speed c. It's ambitious and even makes me wonder about overcoming those disparities between our laws of physics. There's several other issues surfacing as well. I'm going into time reversal, anti-matter, the uncertainty principle and black-holes. All of which smack of our barricades.

Initially, theory still focusses on sub-atomic phenomena. I think it's easier to locate the nonlinear intensity of nuclear forces there. Even so, eventually I'm drawing some correlations between all levels of spacetime and motion. We're just starting the push for conceptual adjustment with quantum theory. The conclusion, by the way, extends the theory of eternal cosmic infinity. Also, from here on in there's diagrams. Someone finally figured out the computer's graphics. So who knows, perhaps the greens or pinks of a squiggle prove as creative as an equation. Then again . . .

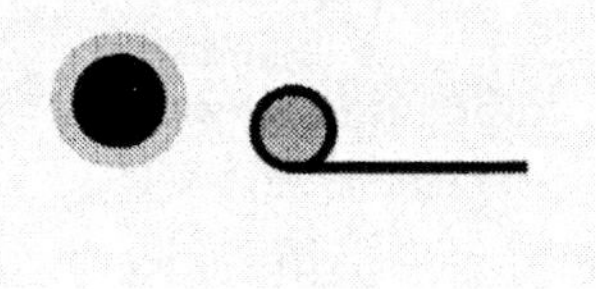

. a way to overcome the finite rationale

In an infinite eternal universe, nothing is meaningless. There's always something. Similarly, beginnings and endings wouldn't be finite. What's so difficult about that? Then we'd come from something and go into something else. You're getting a case for life after death here. Just what that subjectively is though, I'm buggered if I know. All I know is the scientific community resists the idea. Or was it my lousy writing? It could've been the piffling bibliography. More than likely, it didn't make any sense. And there's no evidence supporting these claims. The thing is, applying a definitive beginning and ending is apparently the standard way of rationalising the universe.

Nevertheless, our theories are rebounding from these limits. They express time reversal, anti-matter and uncertainty. The experiments seem to involve quantization that is restrictive by the fixed set of values imposed. Their results seem to concern effects rather than entire causal processes.

We know our laws of physics have gaps in them. And it's difficult to draw cyclical connections through them. At their boundaries our theories bounce backwards, on both the micro and the macro level. We probably need different numerical expressions. Yet we seem to apply the same quantised values to fill in the gaps.

Even if we explore the boundaries of the speed of light we still come up with time reversal stuff. The superluminal activity of the 'tachyon', for example, shows effects preceding causes. There's no way out of it, you think? The categories through which we quantify are subject to proven laws of physics that have a speed and time limit. I know I keep on saying it. But you see it keeps on coming up. We move backwards and forwards for symmetrical validity. Well, I think there should be links through different forms of motion in space and time that could require quantifying. These would involve conversions in motion and continuity not discernible through the present laws of physics.

Gravity on the planetary level apparently shows time going forward. An expanding universe apparently supports this. But it seems we still allow for a time reversal, expecting to detect dark matter to halt the expansion. The way I read it, cosmologists use time reversal through gravity to prove singularities, black holes, worm holes and anti-universes. But why should a contracting phase and the missing matter need a time reversal? Both these issues could arise from the spacetime concept.

Motion, as a missing component inherent within spacetime, should negate its finite aspect. We don't need the beginning then with time. If motion is always there—necessarily—with space and time, bang goes your beginning. I mean you don't get any one and only big bang. As it stands today though, the spacetime concept only seems to recognise motion as a reactionary occurrence. Time curves space, or vice versa, which I understand manifests motion. What's the effect of gravitational mass on spacetime, if not reactionary motion? And I know there's reactionary motion all round the place. That's not the issue. Where do you find the inherent stuff though? Could the integral spin of a proton offer some clues?

Furthermore, the format of motion in spacetime could change through the densities of the four forces. We could get a different combination of spacetime and motion for each force. But I don't see how any of them, no matter how incredibly nonlinear they twist and twine, go backwards in time. Speed c could basically be a conversion factor. If so, it would have great relevance for force interactions. This then is laying the foundations for superluminal activity. And what a jolly back ache it is. I wonder what those ergonomic chairs are like.

It looks like motion doesn't always accord with our forms of measurement in shape and speed. The random element comes in when we try to interpret waves according to the position and velocity of the sub-atomic particle. Which only goes to show the universe at its basic level is incomprehensible to us. Is it because we don't have enough links between motion and space and time to fit the bill? Why do we only have waves or particles anyway? What happens in between and next door? Well, I don't think we're defining its conversion and transformation into matter and energy according to its force differentiation. Meaning that we're not looking at mass and energy according to a systematic and symmetrical differentiation of spacetime and motion into four forces.

How matter and energy transform seems different inside the atomic system. It could more intimately concern the differentiation of spacetime and motion into four forces. We're possibly missing all the dimensions of the sub-atomic particle's raison d'être. We seem to interpret quanta coming from nothing for short intervals. We snap it in at light speed. End of story. So quantum theory closes and seals the circumference for mathematical and quantified theoretical validity. I mean it works. But clearly it limits the rationale.

Where's your spin come spiral relevance? Where's your systematic input? And it's no use reasoning through the gravitational spacetime curvature. That won't tell you nothing inside the atom. But taking the finite beginnings and endings backwards and forwards won't either. They don't cover all the nonlinear possibilities. You're only going through reverse symmetries that way.

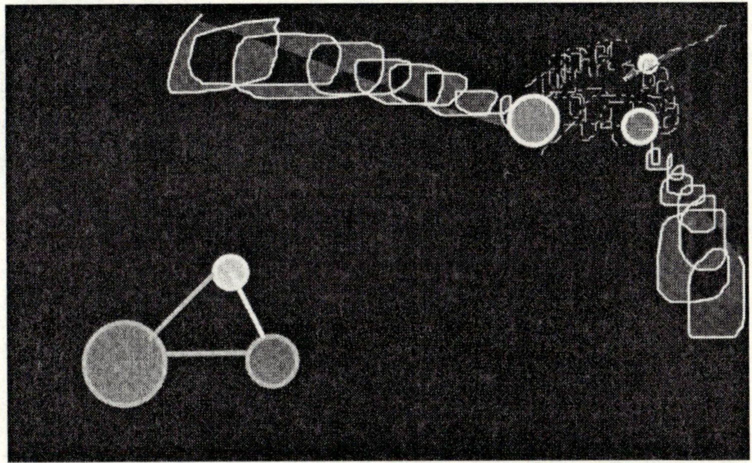

The top set-up in this diagram shows a hypothetical sub-atomic situation where speed c is no barricade. Take it as a symmetrical imposition. We might be looking into the nonlinear dimension here. In the foreground is the standard set-up. Can you see how the effect of motion within space and time figures within the top one? With visible light you don't get it. It's too variable to see all the details. And you're still subject to speed c. But why keep down sizing it through the quantum theory of particle physics? It's like transposing a rainbow into a set of golf balls, teeing off and then trying to calculate the motion with position.

Both general relativity and quantum theory describe discontinuity in speed and mass variables. Perhaps they both use a proven format subject to boundaries. Could they just be quantifying reactionary motion and formation of spacetime at *foreseeable* energy levels and ranges? They show forces becoming particles, which produce more particles, which produce more particles and their flip sides. These quantised packets could express motion and spacetime in reverse. The stuff we're quantising n b1 might be what's rebounding via the wave format at speed c, for instance.

Using our laws of physics, the discrete quantum of spacetime and motion might only calculate through a plane of linearity. That's how it looks to me. But then I'm not up with the latest blasted particle photos. I'm wondering though, whether we capture all the spirals, spinning spheres, waves and orbital interconnections of spacetime and motion in the atom. We sure don't seem to get the forces that way. Especially that is, on the sub-atomic level of nuclear systems. Those forces could be going by way of the inverse symmetry. Take gluons. Look at weak force leptons. QUARKS. Who dares malign all these hallowed little beasties? Unless they improve with age in a bottle, or go with ice and soda water, I'm not interested. Well, O.K. then, so they've got more hard evidence going for them than theory's spirals and spin. Well, O.K. then, how does the lepton figure on a spiral? Does the quark express some strong force spherical spin? As I'm not au fait with them I can't say straight off.

Theoretically, links between the electron and the nucleus could have a quantifiable position subject to speed c. Theoretically, we're rationalising them through a dimension transcending speed c. So the electron might involve more links between the weak force and the nucleus than we pick up through leptons. You can see it in the nonlinearity surpassing a wave and particle. The weak force could get around in spirals for instance. Here we have a discrete differentiation of spacetime and motion feasibly only relevant to the weak force. Theoretically, the weak force has more motion and less spacetime than the wave. In a sense it seems to be elongating some strong force intensity. And that's something that can then transpose into a wave of electromagnetism. I'm even suggesting that it involves a fusion of mass and motion—but more of that later. The thing is; our leptons might not reflect any such values. Leptons could just be linear bits and pieces of spiral that we define by way of speed c. In other words, its spin could come over in fractions according to frequency at present.

Antimatter doesn't seem to have the potential to be faster than light either. This stuff could be your linear flip side of lepton, quark, etc., co-ordinates. The motion, however, of a spiral formation presents a link beyond those quantised levels. It's nonlinear. So it displays both sides and more of the linear plane in its presentation. Position and movement are here taking on a different meaning. The position is subject to the motion through a spiral *formation*. But I'm digressing. Theoretically spirals don't just transpose energies and speed within the atomic nucleus. I think a spiral can dilute the concentrated spherical motion of the strong nuclear force with space and time, hence the different formation. Now there's an unconventional quantification. I'm looking at the antimatter and the matter together through another perspective. And I think you can only do it, spiral wise, if you throw the speed c barrier out the window for a while. Theoretically it's only a wave symmetrical imposition. That's what it's all about. The spiral formation is another one. It too should express an invariant symmetrical structure within mass and energy transformation. You see; these motion formations are supposedly effective within eternal transformation processes. I wrote about them in the last chapter.

But getting back to my digression; position and movement according to a force *formation*. The linear aspect of the wave motion is as far as the laws of physics take us. If so, we're probably not gauging it as a symmetrical junction. It probably comes over more like the end of the line. Evidently, the wave formation is as far as we can go in distinguishing energy from matter, regarding motion. We're at the zero rest stage with it, I mean. Plus your uncertainty principle steps in. It conveniently covers every gap accompanying a previously formulated quantum, or the quandary of movement/position. The thing is; position and movement of subatomic particles could involve the fusion of mass with motion according to the combination of spacetime and motion. The spiral combination is even more nonlinear than wave so would confuse—for us—the position/movement reasoning further. What I'm saying is, the quanta of messenger particles for forces, the uncertainty principle, the linear format, all these things seem to overshadow any motion links of energy transmission that could explain superluminal continuity. They put the blockers on the nonlinear rationale of spacetime *and motion.*

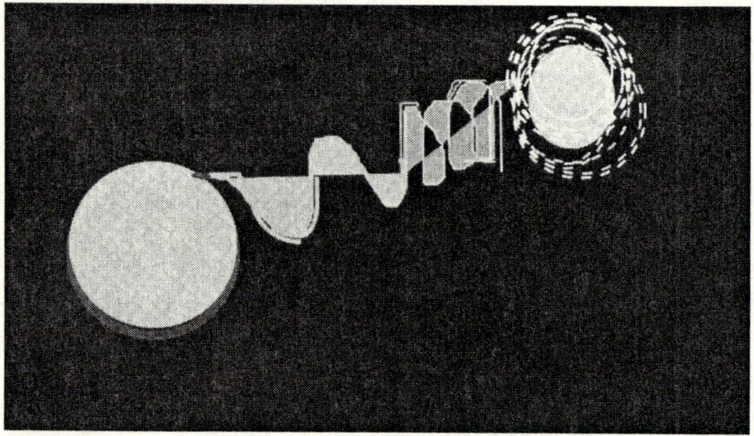

In summary then; we could be quantising cosmic processes that want further quantifying. Theoretically, transformations of mass, energy and motion go beyond the values imposed by the present laws of physics. I think they're subject to an underlying motion within spacetime. And the forces of spacetime, as expressed though motion, might quantify through a formation. For example; orbitals, waves, spirals, and spherical spin could all describe four combinations of spacetime and motion. These could be our basic symmetrical structures of spacetime and motion. They all reflect the nonlinear dimension. Thus far, however, it appears we only quantify the cosmos through the linear dimension. It seems to extend to the speed of light and a wave formation. The nonlinear dimension though, would transcend the spacetime and motion of gravitational general relativity. So addressing it could be taking an unorthodox voyage down a proven cosmic pathway.

Conceivably the intricacies and sensitivities of a system could obscure its causal objective elements. Well, in so far as our powers of detection go. And here I'm referring to the quantum system of sub-atomic phenomena. I'm even wondering that we presume powers of creation within it. Look at our quantum zoo. Inside it could reside the results of our quantising abilities. We seem to be manufacturing the members. Apparently it happens through controlled nuclear reactions and particle accelerators. But take a closer look. These things seem to show effects rather than causal elements. Take the re-activated anti-particles. Are they in their natural state? They're unstable and disintegrate rapidly. And producing them destroys their natural symmetry. Are we just devising some sub-atomic connections of a linear plane? Perhaps we're just piecing together quantised effects rather than addressing causal processes. So what if getting them involves all that sophisticated technology. And maybe I don't understand the tiniest component of a particle accelerator. But what's the real point of a quark if only the particle accelerator can determine it? Maybe we're just creating a load of inexplicable debris. Maybe that's what this is. Well, that does it. I'm knocking off for the day.

Within a system considered basically random our subjective element may have grown out of proportion. Perhaps we're 'creating' and studying the effects of nuclear debris, rather than understanding entire processes here. Why do our measurements of waves and particles exclude each other? What are the theories and objectives preceding our measurements of them anyway? Like Einstein, I think God doesn't play dice. I think he may have dealt us some cards, but we've yet to understand the game. Indeed, the standard claim is that nature at some level is chaotic, and indeterminate.

And that's coming from us. We're the ones who still haven't figured out continuity on both the micro and the macro level. So that means there is none?

. spacetime and motion: a new conceptual basis

Einstein apparently rationalised motion *subject to* mass and energy. I don't know if he took motion *into* mass or energy with his calculations. I don't think so, because the concept of spacetime doesn't look to be a dynamic entity within itself. Look, maybe I'm wrong. Again, I'm no authority. Nonetheless, there's nowhere that I've read anything that suggests other wise. With all due respect to the genius and brilliance of general relativity, I don't think it explains all the cosmos. Mind you; it's certainly more salient and translates into reality better than this theory. Einstein calculated everything. I'm not putting two buttons together for any of this. (Which is probably why I can go the eternal distance.) But what I'm getting at is; general relativity seems to reveal structural laws dealing with *reactionary motion,* or a certain force of it according to space and time. It doesn't explain the 'spontaneous' energy inside atoms, for instance. There could be causal processes of infinity these laws don't describe.

You know; I don't think you can have infinity without eternity anyway. Well, not in it's full blooming brilliance. And those causal processes I keep alluding to, could derive through that greater dimension. Putting motion into the conceptual basis of spacetime describes that dimension. It becomes dynamic. Your space doesn't have to start with time, produce motion, then stop at speed c. Instead of voids, we have spacetime and motion through which to deal. There's just various configurations of this ensemble. So what may seem like a void is not a void—on any level. Everything doesn't need to go into a packet of spacetime *in* motion, for meaning. There would always be some form of motion inside the particle, or the void, this way. And the particle would always have some form of continuity. So would the void.

Besides out of a pub, this is how you get your nonlinear movement. The motion becomes subject to a formation which can then reflect all the dynamics of a force. And as we have four forces, we should have four formations for them. That's four different combinations of spacetime and motion. I'm calling it the metaphysical grid. A spiral motion for instance could relate to the weak nuclear force. It should reflect a specific formation in space and time. Theoretically, the spiral *formation* could also be an elongated encoded remnant of the spherical situation of strong force activity.

Now if motion is inherent in spacetime, the dynamic element of the cosmos could transcend any one particular force, in one form or another. You're getting a more comprehensive spacetime continuum this way. This is aligning us to the bigger picture. We can go beyond the big bang with it. Rather, a primal cause becomes a systematic event. Your big bang is subject to the forces of spacetime *and motion.* Simple. Any such intrinsic causal connection, however, doesn't seem apparent in either quantum theory or general relativity. Look at the general gravitational spacetime curvature. Does it describe an unfolding event towards spacetime only in terms of the mass of some general level? On the micro level, we could be looking at a system where motion is more condensed within spacetime than that of the greater gravitational level. And if so, treating it as reactionary with spacetime might prove uncertain.

In the diagram below I've drawn part of a hypo sub-atomic system. There's a spiral of spacetime and motion transpiring out of a pink strong force formation. It's linking up with a green orbital corridor of spacetime and motion to reveal an electron. And where are the electromagnetic waves, you may ask? Where indeed. Well, they're off the page and out of my brain. Perhaps they'll show up later on.

1378-RUSS

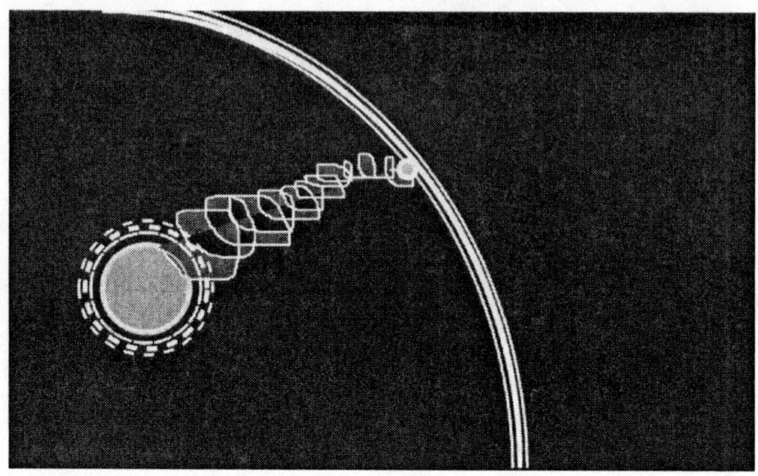

This electron set-up could be subject to a host of nonlinear inter-actions. Theoretically its particle formation and/or wave movement exceeds the values of the linear format. They quantify into spirals, for example. And spirals, on this level derive more closely by way of other force formations. We're ostensibly looking at a com-pounded, nonlinear dynamic arrangement within the atom here. Give or take the waves, we're getting the whole four force system. Outside it, you're only getting a gravitational spacetime curvature by way of below speed c motion. You're not getting the general level four force system. One demarcation between spacetime and motion doesn't cover the entire system. It only shows us what we can see through the greater plane of linearity afforded by its size. The size that is, of the bigger nonlinear picture. For the four gen-eral level sized forces to figure systematically you'd be looking past the big bang beginning and the black hole come quasar situation. Motion within the spacetime concept provides the nonlinear di-mension. Theoretically it surpasses our quantised values of the four forces. And this dimension relates to both the world of general relativity and the quantum domain of atomic systems.

. inherent motion in spacetime and the four forces

With motion you have a dynamic set-up from the word go. It's a simple formula through which to rationalise the cosmos. The idea is that all the forces are manifestations of motion, space and time. Or one force if you like. They're all apparently responsible for life. Spacetime per se does bugger all. Sure, in generating motion it manifests a whole range of processes. But the gravitational general relativity through a spacetime curvature is clearly not the only thing going on. That could just be one facet of a system

With the spacetime concept and unfolding motion I don't see how you can reason out the systematic process. You can't see why the space curvature begins with time. You only see the effects of the event. With inherent motion though, you're making spacetime dynamic despite gravity's relativity. We're opening a window of reasoning here. The four forces all bask in the one spectrum through it. Spacetime and motion means they manifest from one basis. They all grow out of the same ground. It's the dynamic dimension. And it divvies up into four forces. I'm just suggesting that we get different concentrations of it. So each combination of spacetime and motion can produce a formation, thus defining the energy and behaviour of four forces. It seems to be a symmetrical occurrence. One that realises balance throughout every entity within the cosmos. Without nuclear forces you wouldn't have gravity, in other words. The differentiation of spacetime and motion into forces is a systematic process.

Within the atomic system, it's easier to discern this. The atomic shell seems to encapsulate the interactions of four force formations. I'm wondering that it shows them inter-linking in a symmetrical systematic set-up of spacetime and motion. Whilst at that level interactions are more intimate and more than likely, interwoven. You see; inherent motion within the force could provoke another force to act on it. Now that's despite the force itself only acting according to its motion combination—according to its niche in the metaphysical grid. Inside the atomic shell these forces seem to fire away together. As a system you're getting manifold interactions here. The sub-atomic weak nuclear force apparently contributes to electromagnetic power instantaneously. Once outside the shell though, these spirals and waves wouldn't do a lot together. They probably don't even know each other out there. But they'd know the stability of greater gravity below speed c. They're no longer subject to the atomic *system* out there. I reckon it's a different ball game out there. It's another system of force formations, where the atom *per se* has more meaning. But I'm jumping the gun on that one.

The thing here is; inside the atom spirals of weak force spacetime and motion seem to contribute to electron activity and the wave formation of electromagnetism. At the same time, spirals appear to develop out of the strong force of spherical spin. Here then, the weak force is *acting* upon the strong force motion. And the strong force is *acting* effectively in its spherical set-up. Their system's symmetry depends on their balancing act. Together they are divvying up the atomic spacetime and motion into symmetrical forces of action. Look, on any level the strong force manifests through the minimal space and time quota with maximum motion to realise a spherical formation. It's just that here inside the atom you're getting the systematic input more clearly. Indeed, strong force spacetime could so diminish here according to the symmetry of the four force system, you can decipher it as a nonlinear fusion of mass with motion. And here inside the atom, it could be what binds sub-atomic particles together. But that's also jumping the gun.

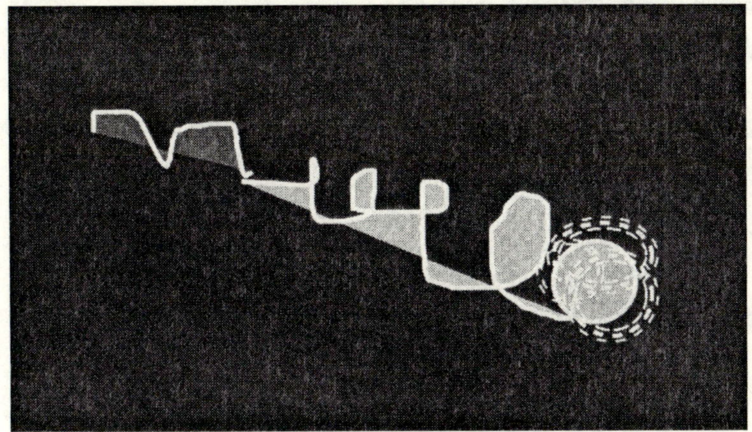

In the above diagram, waves, spirals and spherical spin are all equal manifestations of spacetime and motion. The spiral above is diluting intense strong force by way of the greater space and time and less motion of its formation. The wave does the same thing with the spiral.

In this convoluted fashion, the sub-atomic weak force should only act within the nucleus, but its motion you see, once formed, transcends the nucleus through spirals. It can therefore *interact* beyond it. Initially the weak force might change the nuclear spin by way of releasing some of its angular momentum. The spiral formations are able to carry that momentum in various directions beyond the nucleus. And other stuff can then act upon them. Electromagnetism could start happening through them. Bizaar. Anyway our sub-atomic particles for me, are a tangible effect of all that. Their energy and mass derive through the interactions of a force, and indeed theoretically reflect its definitive motion, space and time combination. A proton, for example, could be the physical realisation of strong force activity.

Here inside the atom our forces could define each other more so than on a greater level. Each formation acts through different portions of space, time and motion. Even though, theoretically they would pan out in some balanced sense as the same. Where one has more motion it has less space and time, and so forth. The wave has more than the spiral. So the extra space and time can dilute intense nonlinear motion of say, the strong force. And so it goes down the line. What we perceive then, as attraction and repulsion, might still tally up with the two way movement, and spacetime and motion content of a wave motion. Negative and positive forces could fashion out a wave movement via the symmetry of the system. But I'll explore all that further down the track. And if I'm stating the obvious, don't be under whelmed. It means theory could be aligning with some facts. Far out.

Whatever. The thing is; only at the level of an orbital formation can we quantify physical phenomena. And that seems to further confuse sub-atomic stuff for us. Remember the wave/particle duality and uncertainty principle? I'm not about to forget it. It's a beacon of encouragement. I think it reflects our inability to discern the close knit interactions of a compound system of four forces in the atom. And that could favorably reflect upon my inability to lucidly explain anything. In other words, we might also be getting a gravitational orbital corridor of space and time (and minimal motion) inside the atomic shell. Call it a micro force formation system. No equations, you see. I've got to call it something. And besides, I can say these things because I'm not staking an entire academic career on it. What have I got to lose; what reputation, what career? I'm living in the woods on the dole, baby. This is an unknown with unimpressive qualifications. I can only dazzle with audacity. This is what it's all about, after all. How else are we going to get beyond the speed of light? Einstein had a physically different brain from us. I saw the documentary. It's dissected in a bottle some place. And probably the only physical thing this brain has in common with that one is the pickling business. Anyway, I'm still going into all these subjects later on. We're having systems inside systems all for the sake of symmetry, right up to the cyclical level.

. endnotes

nb 1 *quantise*: to restrict to one set of values.

quantify: to express or discover the quantity of something

ch.3

force formations

gravity: orbitals
electromagnetism: waves
weak nuclear: spirals
strong nuclear: spherical spin

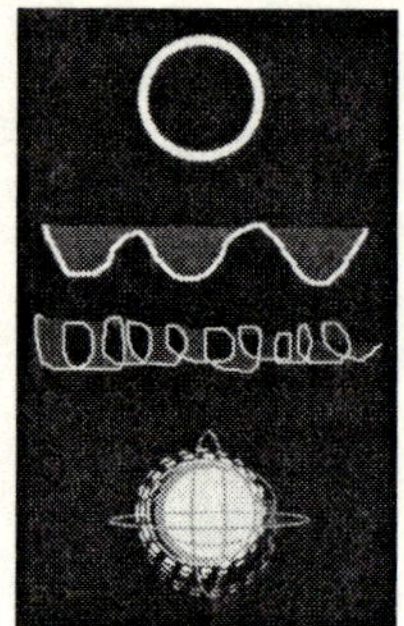

A simple way to distinguish the four forces within spacetime and motion is through a formation. It makes it easier for working things out later on, down some complex line. That's the only way you'll get them from me. I'm calculating nothing. These symbols render further hypothesising a singe. Even I can tell that the different combinations of spacetime and motion should manifest different formations. The weak nuclear force, for instance, comprising of less space and time and more motion than either electromagnetism or gravity, would display a more nonlinear set-up. The motion would become more nonlinear by condensing within a tighter frame. Ergo we have the spiral formation of weak nuclear force activity.

Try the same logic with gravity. It should create the largely linear formation of orbital motion. Ostensibly, an orbital only has the one-way movement of massive attraction. But the point here is this one has the most space and time and least amount of motion. So fathoming any nonlinear fusion of motion into mass through it is near impossible. Besides, you'd have to consider the scale of the system. Sub-atomically your mass-motion fusion is more systematic. Whereas the linearity of directional motion in general level gravity, that's jolly obvious in the bigger orbital, seems to arise in reaction to mass rather than exist as an inherent element of spacetime. Nevertheless, an orbit should still comprise of inherent motion. But like those elusive little gravitons, pinning some down could be pushing it. Nevertheless, I think we do it in about chapter 8.

And then electromagnetism, through attraction and repulsion, seems to curve spacetime and motion up and down. Even so, this two-way action still seems pretty linear. You don't have any spin here, only frequency. As such, electromagnetic activity appears to define phenomena in the linear format of gravity. Stuff doesn't twist around on itself. You're not getting nonlinear mass-motion fusion. You only get that same hollow curvature heading inevitably forward. We might however, find out otherwise, in light of its dual nature. The wave/particle one that is. But again, this would also be subject to the systematic interaction of forces and their scaled rate of activity.

Both the nuclear forces have more motion and less space and time than the other two. Theoretically, these go right into the nonlinear heart of superluminal activity. Planck would probably disagree. Well, for that matter, most eminent scientists would probably disagree. Who's denying it, I'm evaluating Plank's discretion. According to theory here forces only figure on the same scale through a nonlinear picture that includes Plank's gaps. And this nonlinear dimension covers the linearity of orbitals and waves. The spiral formation, however, carries spin. And this signifies a nonlinear fusion between mass and motion—which signifies superluminal activity. Space and time being so scarce the motion seems to par boil it. Well, twist it around upon itself into a tiny weighty entity. How many analogies do we need? It makes for power and the strong nuclear force has the most of it. It packs motion into a tight spinning ball. I call it solid rotation. So, Plank, I'm still with you in a way, because we've fashioned an individual quantum domain.

But those discrete sub-atomic orbital levels would always have a force formation separating them, in some manner. So I'm not with you all the way. The remaining forces and their systematric symmetry covers your discretion.

Interestingly, the weak nuclear force seems to carry the imprint of nonlinear spin. Spirals after all, are the supposed remnants of strong nuclear activity. On that basis, their formation could show a directional polarisation with the imprint of spherical spin. Spirals could describe some highly nonlinear angular momentum in transformation. Maybe the particle imprint of spherical spin permanently encodes them. There's your twist in the spiral. Either way, I think spirals require deeper consideration and further evaluation. Their mysteries seemingly illuminate those barricades to the land beyond light speed. Take the ones resulting from particle accelerator collisions, for instance. They still appear to quantise along the lines of the time reversal of speed c limits.

On face value, even photons look like linear interpretations. For all I know, physicists find this quantum of light through a wave frequency rather than a spin factor and the formation of mass. But with zero rest mass, I reckon the photon is expressing the nonlinear set-up of a mass-motion fusion. And from what I can gather, photon activity should also include the radioactivity of a spiral formation. They'd be subject to spirals from the strong nuclear spherical spin. Although travelling at speed c, they wouldn't be *acting* within the spiral format, but only through the wave formation. Now I want to get this right. So if I sound like a pedant, it's for that reason. *Theoretically,* the spirals are only *interacting* once they get beyond the nucleus. With the photon, we're measuring photon activity. We're measuring an up and down movement according to a linear format below speed c. We're only judging it by the wave. But photons, like other sub-atomic particles, would affiliate with a full circle of spinning reverberation. All such phenomena should reflect a close interaction and subjection to a *system* of forces.

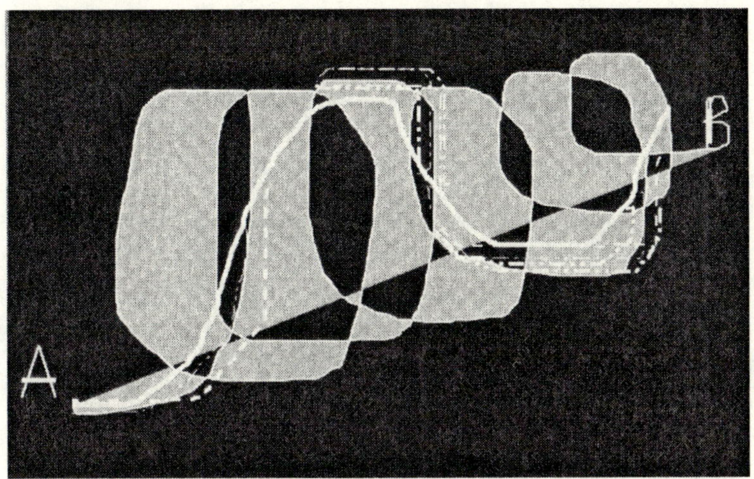

In that diagram a wave superimposes on a spiral. The wave is moving at the speed of light. Now both the spiral and the wave arrive at point B at the same time, having departed point A together. So If I was riding on the spiral formation I'd be going faster than if I was on the wave. The point is however; we wouldn't detect the spiral motion between two defining points of waves because it would be superluminal. Another way of looking at this is by the curves. You could say that the more curves involved in the formation the more motion. Even going by the spacetime concept we only have motion where time curves space. The difference here though is that the whole three are interdependent. The formation just reflects the different combinations.

Finally, the strong nuclear force describes a particle formation. It looks like a sphere, and is theoretically spinning spacetime and motion. Now I think the spin forms a mass of buried treasure. This activity could well surpass our present laws of physics. Can quantum mechanics, for example, pin point the axis of spin of a sub-atomic particle? Or does it, rather, describe what the particle looks like from different directions? Could it draw connections between an elongated spiral trajectory and the proton?

Theoretically, spherically spinning rotation forms the heart of energy levels that transform through other force formations. You've got an amazing amount of massive increase inside this tiny formation. As space and time decrease, motion increases forming the microcosm of a massive nonlinear fusion. Look at the mass of a proton in it's tiny set-up, compared to a wave of photon. And here's a claim to render all previous ones pedestrian. Going by Einstein's equation, the strong force formation motion could be a nonlinear fusion of speed c squared essence. Perhaps spherically spinning strong nuclear force energy can amount to mass moving at speed c squared. That's my buried treasure. Give a dog a bone . . . Einstein, and if I dig it up and unravel it, it might power us into kingdom come.

. an atomic system of forces?

Our chances of uncovering any such power probably remain pretty slim, if not impossible, according to the present laws of physics. It's apparently impossible to distinguish the effect of gravity from non uniform motion for starters. And I'm talking about a systematic process here, between 4 forces that involves both. I'm still pointing a finger at the linear dimension. Applying it to gravity could prevent reasoning motion multilaterally. Gravitational motion won't translate into the condensed framework of nuclear activity. So we lose the continuity of a systematic set-up. Rather, continuity and motion going beyond the lateral simplicity of the wave function becomes an approximation. And forces and motion, quantised as matter with mass, only express energies up to speed c.

In the diagram below, a sub-atomic particle is moving in a pink wave. But I think with the pink we're looking at the linear dimensions of a nonlinear formation. Now that's something that would have an element of mass-motion fusion. See how the greeny-blue definition of a position could negate the motion. To some degree the position is the motion. And it could appear more so with a spiral. Because here also, is the impossibility of defining mass at speed c and beyond. There is either position or motion. With spirals you're beyond speed c and it's even more diffused. Waves, spirals or spin, you'd have to look at the sub-atomic set-up through it's systematic rate of force formation interactions to get the complete picture. And that means entering the hinterland beyond light speed within a nonlinear dimension.

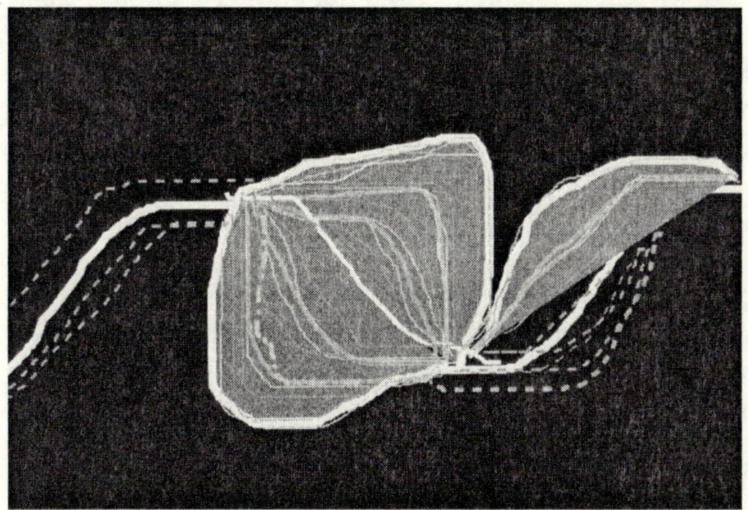

You see; I'm trying to describe matter and energy transformation through the spectrum of force formations. I think it could reveal phenomena we don't normally see. The gluon and quark could take on new meanings. Perhaps they're our linear land marks. Anyway, theoretically [nb1] these interactions describe the continuity a linear format disqualifies. They could reveal what goes on behind the closed doors of gravitational general relativity. We might even be able to decipher contrasts between macro and micro systems of forces.

Our force formations should describe combinations of spacetime and motion transforming into mass and energy by way of their interactions. For example, the electron might be subject to an orbital corridor of spacetime and motion, as well as the spiral input. It's energy could derive by way of different forces of spacetime and motion interacting. Indeed, the electron formation seems to contribute to the electromagnetic wave formation. We could, as such, be addressing a systematic occurrence. And the key to it all is symmetry. These things should interact symmetrically through four forces to realise a system.

I'm therefore proposing that the atomic system involves connections between motion and mass/energy conversions through *four* forces. That's 4 atomic sized forces, ergo the atomic system. And I don't think we generally reason sub-atomic phenomena that way. Gravity always seems irrelevantly outside the atomic shell. And maybe we're only getting one version of nuclear activity. If I'm right about this, nuclear activity would also figure somewhere on a grand scale along with the gravity of general relativity. The big bang activity preceding gravitational expansion might be a general scale nuclear event. Likewise, a gravitational force should also figure within a small scale system along side small scale nuclear activity. For us, electromagnetism seems sandwiched between two levels. And it shouldn't be. But that's another symmetrical story. There's a deeper transformation picture going on here both inside and outside the atom. One that shows interactions between strong and weak nuclear forces, as well as orbitals and waves.

Proving the scenario is practically impossibility because no-one's detecting stuff above speed c—*inside* or outside the jolly atom. Just because we can't get around the galaxies beyond light speed, you'd reckon we'd at least be able to measure things inside a piffling atom at those speeds. You'd think. Then again, look at those computer viruses. How small are they? Imagine a superluminal one zapping around your networks. For sure, some insidious mind here would strum up a superluminal nasty, putting a spanner in the works for fun and vengeance. Besides that, maybe we don't detect weak force spirals and high velocity because quantum theory quantifies all the forces into packets of energy below or at speed c. They appear to materialise into the appropriate mass, energy, spin and quantum numbers according to the laws of physics. Just like your carpet though, these particles look like they're in a linear format. And no matter how many you quantise you wouldn't encompass everything with them. Their gaps would still involve spacetime and motion.

And if that hasn't given you mental fatigue, here's some more. Sub-atomic phenomena, that's radioactivity too, should involve nuclear mass, right? I'm reasoning nuclear mass as a nonlinear mass-motion fusion. This then transforms via the elongation and interactions of released condensed spacetime and motion. You get spirals of the stuff transpiring into waves, etc. The density of the spiral formation ostensibly reflects the remnant energy of spinning momentum. There could be loads of it coming from the tightly bound spherical spinning set-up. Any imbalance building up within the spherical spin would churn it out. I mean, with motion in the works you're always subject to change. There's always the potential for symmetry going off kilter. But when, how, and why the strong force does this is off the page for now. That probably comes up in the chapter on cycles and symmetry. What I'm getting at here, is the *systematic* response to the provocation of strong force build up. Perhaps we can measure it through these force formations and their interactions inside the atom.

That equation, $E=mc2$, could co-ordinate with each force you know. Look, on the macro level and below speed c, motion and continuity make sense. Orbital mechanics and orbital motion signify our space travel, for heavens sake. But impressive as it surely is, it doesn't enthuse me much. Gravity and mass are useless to us for the greater part of space exploration. We define them uncertainly with higher energy, thus limiting space travel to orbital motion. Here we go, silently gliding through space with the momentum of a toadstool. Well, O.K. then so I'm not au fait with the ins and outs of Voyager's mechanisms. But I wouldn't hold my breath anticipating any inter-galactic insights in this lifetime, not by the way we're going. You might as well roller blade across the Goby desert.

We should be dealing with 2 *systems* of forces, if not more. You've got your generally relative one and sub-systems within it. And your generally relative one should still only be one of many anyway. However, our ability to master gravity and mass on the macro level seems to over rule everything. Because gravity figures as far as we can see we take at as universally applicable. Even the thought of higher energies, and speeds of mass through spiral motion and nuclear spherical spin, seem lost. Our ability to fathom nuclear forces on the sub-atomic level relegates that to meaningless discordinance.

What's so inconceivable about motion and continuity, between and within, all levels of phenomena? How can one system exist in isolation to another? Clearly they don't. Why should the order we determine not be part of a systematic occurrence anyway? Clearly our present laws have limits. Are we so full of our achievements that we can't see beyond them? I wonder what Newton would come up with using today's high tech info and equipment. Why can't we figure out mass and gravity at higher energy levels, or inside the atom? We detect the signals but we don't draw connections. Outside the atomic nucleus, for instance, the neutron decays into an electron spiral. And within the atomic shell quantised particles behave as if they have angular momentum. But quantum theory neither draws correlations with the electron spiral motion nor defines certain spin. It looks that way, anyway.

Makes you wonder there's room for further thinking. Too bad my brain's not up to it. I'm going dizzy at the gills. Wouldn't it be nice to stretch morning tea into lunch into afternoon siesta into happy hour? Now there's a thought. Who knows if it can be done. Who knows, maybe nonlinear motion—the formation, can illustrate force interactions on all levels. Maybe they can even evaluate the undetected stuff our present laws predict. Based on the inherent motion within spacetime, these force formations should cover the gaps left by quantum theory and general relativity. Look at the broken remains of spiral formations, deriving through particle bombardments. How can we gauge the full atomic picture through them? Why should the mother of all domains, the atom, only fit into a linear level jig-saw puzzle?

Theoretically, interactions of motion formations describe changing energy levels. And that's supposedly on any scale. I'm taking these force formations as conversion agents. They should impose the invariance of symmetry throughout all transformations. Furthermore, the velocity differences of each formation would be an integral part of this. Velocity is a part of any force and that should translate to any mass/energy transition. I thr force of an orbital formation has velocities below speed c. A wave formation travels up to and at speed c, and a spiral formation transcends speed c. Finally, a spherical spin would have the motion of speed c squared. That's the metaphysical grid again. Moreover, these four combinations of spacetime and motion should realise different transformations of matter and energy according to Einstein's E=mc2. It should apply to any scale of a force formation system. That's the nonlinear dimension for you.

The differentiation of spacetime and motion into four forces is supposedly a symmetrical event. So the velocity of one force counter balances that of another, in terms of their space and time. It makes for a system. There's your stellar systems with solar spin and energy and mass transforming into planets around it. There's your galaxies happening. They're all symmetrical. Stellar systems, galaxies—they all maintain a sense of balance, don't they? And overriding the lot is your general level one. In the smallest system, say the sub-atomic one, the forces theoretically interact most in tandem. But as you go up the scale, they seem to gradually take precedence over each other until on the general level they individually predominate over the whole bifurcating heirarchy.

Take the steady state of sub-atomic stuff in orbit, for instance. Electrons could arise through the interaction of a wave formation as well as a spiral. The wave could clarify the vertical distinctions between compacted and elongated spiral formations. Are you with me, Plankites? That means the electromagnetic wave interaction could stabilise the higher velocities of intensified spacetime and motion. But it also accords orbital distances. So here's your mini gravity coming into play. Indeed, looking at it systematically, the position and velocity of sub-atomic phenomena might make better simultaneous sense. It could appear less uncertain than we know it. Rather, it could just be subject to a more sophisticated energy and mass conversion. It's another way of looking at things. And it's all based on symmetry.

Without getting too technical here, let's take electrons systematically. Starting the process of atomic change could be weak force action within the nucleus. Subsequent spirals could transform excess nonlinear mass-motion outside the strong force spherical spin. They're theoretically diluting it with more spacetime. It's maintaining spherical equilibrium and producing a form of energy. Then you have your orbital corridors to further regulate the energy transfer. Those formations epitomise the systematic process, for me. They clarify discrete distances through their own manifestation from other forces. I'm going so far as to suggest that these things are an effect of positive and negative interaction. In other words, your wave formation, the interactions that manifest it, also seem to contribute to an orbital formation. Using force formations you can see that the orbit regulates spiral motion below speed c. We're clarifying a corridor for the electron spiral position. And then you have your electromagnetism happening. The thing is; these interactions of force formations, should show mass and energy transforming inside the atom. Our photon orbital transfers, on the other hand, might only describe certain segments and effects of the process. I think they spotlight an orbital position more so than the electron's profile.

And what about random spirals, you may ask. They're outside the system. What about neutrinos, etc.? Could they be evidence of what happens to spiralised mass-motion release when it doesn't translate through the atomic system? Are they a consequence of a symmetrical imposition? Only so much spacetime and motion can transform into only so much energy and mass according to these formations. Maybe the neutrino is our excess of mass-motion build up. So maybe it can't transform through the system. So maybe it becomes external to maintain internal symmetry. Then it's subject to another system, which then provokes some indirect response. And that's as far as I'm going down that road. Otherwise we could end up with the effective baloney of a cosmic butterfly.

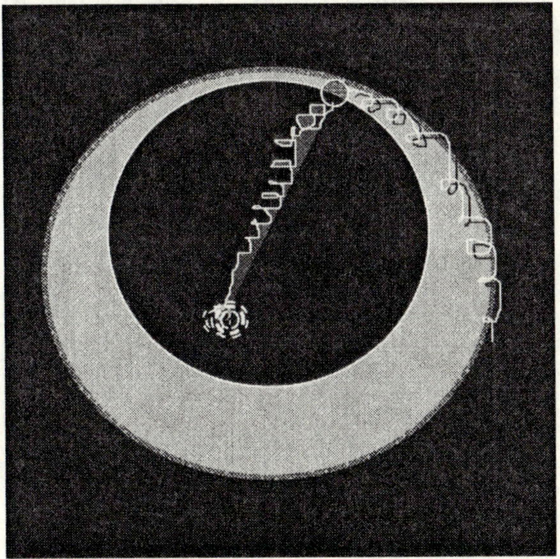

Physicists know what orbit the electron's in by its energy and frequency according to waves. But it may be possible to calculate a value for electron energy and its position through these motion formations and what they amount to in terms of spacetime and motion. Suffice to say; we'd need Einstein's equation and some strange symbols for assistance. Failing that, here's a diagram of what I'm proposing. It's basic, with more gaps than quantum's theory in its logic, probably. Even so, those motion formations could distinguish the effect of gravity from non-uniform motion— all the better when Reubens here works out the electromagnetic presence. The waves never seem to make it to the drawing board. Can you see the motion formations of the weak nuclear force interacting with, and transcending the strong nuclear force, of say the proton? Spiralling around the blue gravitational orbit is an electron empowered by green weak force radiation emission.

Interestingly, spiral formation could figure in both contracting and expanding states of matter transformation. It looks as if it can transfer stuff more rapidly than electromagnetic wave motion, and more intricately and nonlinearly than orbital motion. Yet we know little about the weak force beyond its actions within the nucleus. But look around the place. Spiral formations are just as prominent on the grand scale as the atomic one. They range up to galactic disc formation and down to sub-atomic decay. I saw the latter in bubble chamber pictures.

Spirals come over as the initial transforming agent within the atom. Ostensibly other forces then act upon them. My theoretical position here is that these force formations, particularly the spiral, seem to act in tandem with other forces, more so within the atom than on other levels. And the challenge remains to observe it. We'd be looking at the transformation of high energies. Remember though; we're only doing it through the motion *formations* of four forces. Then again, that's *inside* the atom. So how do see all the action inside an atom? How would I know? I'm not here to invent the gadget.

The interactions of these force formations could show associations between condensed spacetime and motion, and momentum and angular momentum, you know. That's when we're approaching superluminal wonderland. The mass of the strong nuclear force is supposedly a small spatial area within the atomic shell. It should be the most concentrated form of motion in the smallest piece of space and time that you can get. This strong force mass is theoretically, not only a condensed form of spacetime, but also motion. For me, it's the epitome of nonlinear dynamics. The mechanics of this force and how it generates such power are what I'm trying to unravel.

So I'm going deeper into theory with this. There's more of that later on. The formation of stong force spherical spin within the nucleus could have a quantifiable particle formation, or mass. I'm calling it a mass-motion fusion. As the highest concentration of mass and so potential energy, I think it equals the energy that transforms into mass at speed c squared. Closest in helping this transformation, that I can gauge, is the interaction of weak force spirals.

On the grand scale, a state of spherical spin may not represent such condensed spacetime and motion. Without going into all the technicalities however, I think the formula still applies. Yes, clearly I don't know them, but that's not the point. Any idiot can draw a correlation here. We've got the earth going around the sun, in it's solar system. Let's say the sun is the spherical set-up of strong force formation on the stellar level. And the sun, for all we know, could possess the energies to blast Sooty in and out of the galaxy in next to no time, at speed c squared. Then on the galactic scale you have your spiral galaxies and a nucleus set-up within them. A high powered quasar, you think? Taking this to the maximum, our cycle could expand out of, and contract into, a big-bang of nuclear spherical spin. And on that note, I'm calling it a day. We've gone way passed the subject matter of atomic systems. If you're still curious about it, well, it figures further in the book, along with everything else.

. *spirals in the superlumnal pond*

Everything could, to some degree, be nonlinear. The linear could be a measure of what on a greater scale is nonlinear. Are we addressing just a slice of reality? The general relativity of gravity could be a phase within the bigger picture of a cycle. Can the spacetime and motion combination that factors up to speed c cover all the ins and outs of eternity and infinity? The data before us signifies otherwise. Look at your black hole situation. Not until we either reach farther out than our present expansion phase, or take the atom for the system it is, might the nonlinear dimensions become apparent, however. And that, in a nut shell, is theoretically why the nuclear rationale doesn't align with general relativity. How can you go tapping into the cosy little world of our nonlinear micro system with the linear logistics of macro gravitational general relativity? There's another way according to this theory. Force interactions pertaining to the atomic system, transcend the speed c pattern of information that defines the world at large. They involve the whole pie. You get all the cream and apple as well as the icing sugar on top, plus the pastry with them. Too bad we're only detecting nuclear forces and electromagnetism at speed c. So we don't get the spirals. The quantised photon steps in where a superluminal spiral formation might occur in the wavelength rationale.

Perhaps an electron and its necessary links to photons indicate a spiral superluminal potential. The electron could be networking between forces inside the atom. Sub-atomically you don't seem to get the general prevalence of any one force in particular. Not like you do when you step outside it. Perhaps the electro-weak theory covers that. Not that I know anything about it. All I know is the weak force makes more sense to scientists inside the nucleus. The point however is this; through a nonlinear dimension spirals of weak nuclear force could figure in both expansion and contraction of matter. Your nonlinear dimension shows how forces can interact through different formations, whereas the linear one only stretches to the gravitational spacetime curvature. So I'm going into this dimension of spirals. Sub-atomically on the one hand, the spiral relieves the burden of imbalance within the spherical spinning formation. On the other, it can boost the electromagnetic set-up. That being the case, weak force spirals could present ways around those impossible superluminal mass increases present theories suggest. Perhaps weak force spirals can control mass increases *and* defy the luminal limits of a wave.

I reckon the interactions of spiral formations can overcome the linear problems concerning mass. They seem to coordinate the higher energies of nuclear forces beyond longitudes and latitudes. Weak force interactions possibly transfer momentum while counteracting mass increase. They might even translate the condensation of spacetime and motion, according to nuclear spin and mass formation, into the energy of propulsion. What a boon if they do. But who knows how weak force spirals function outside the atom, let alone the nucleus? Where's the nonlinear dimension outside the atom for them to figure?

Don't get excited. The weak force only acts within the nucleus anyway; remember. It only acts through spherical spin. Theoretically however, spirals can be *acted upon* outside the nucleus. It's part of the atomic *system*. And theory here, says we have macro systems of force formations as well. Ones that figure in the greater nonlinear dimension of spacetime and motion right up to the cyclical level. But these ones, apparently, don't function in unison the same way our micro ones do. Even so, your weak force might remain closely attached to strong force spherical spin on any level. I can't really fathom a general relativity of spirals? Is it a quasar or is it an accretion disc? It's probably black hole affiliation, so I'd go for the latter. Whatever it is though, it's not happening within our visible range. But it should function in proximity to your strong nuclear happening. A cyclical contraction phase could realise an accretion disc of large scale weak force activity, for instance. That's macro spiralised combinations of spacetime and motion. And this could happen into the black-hole-come-big-bang of predominating strong force spin. That means a large whack of the cosmos goes under the general relativity of spherically spinning spacetime and motion. A microcosmic event, no less. What a leap of the imagination. I'm on a roll here. Look, spiralised mechanisms may usher in our arrival on the cosmic scene. If we can figure them out with a gravitational set-up and a strong force situation, we've reached non-linear land for sure. And that's not it below. That's a space ship.

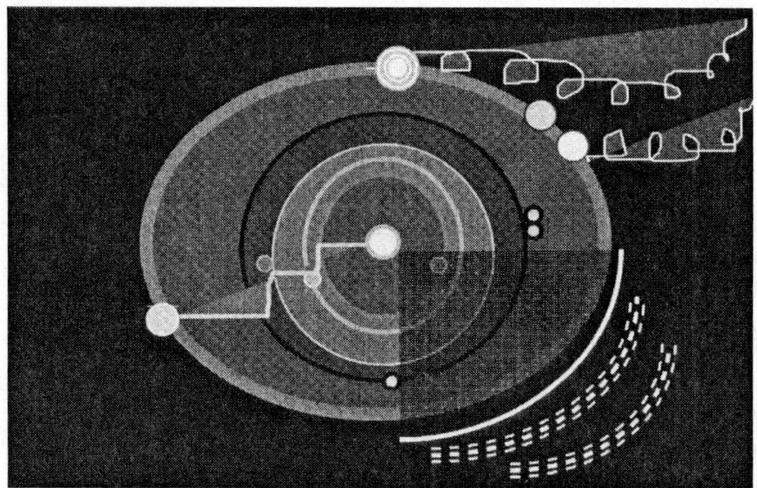

The ultimate idea is to manoeuvre force formations. Imagine a space-ship where your proton of nuclear power has external access, and you've got a gravitational protective shell inside. Then you're not subject to the weightless nausea of Apollo somersaults. Mind you, how a proton functions on the outer gravitational layer is still a large leap of creative indulgence. Even so, spiralising angular momentum might generate power as well as control mass. I mean you wouldn't want to be transforming along the way. So you'd need your gravitational shell for more reasons than nausea. And I wonder how you transpose an electron's orbital? Who knows.

In the next chapters I'm looking at the electromagnetic set-up inside the atom—as a charge system. Even in Tinkerbell's fairy land here one can only ignore electromagnetism for so long. Waves might clarify whether a micro gravitational force goes hand in hand with nuclear forces or not. I want to find out how those orbitals function and stabilise electron energy. Just having *a feeling* about it wouldn't raise a mushroom in bull—well you get my meaning. The atomic forces should closely interact in unison. An imbalance in proton spin might trigger off a charge system of electrons, thereby fashioning a wave formation. It's your forceful symmetrical differentiation of spacetime and motion inside the atom. It might hold clues for superluminal travel. It might not.

. micro-macro correlations of force formations

On the small scale, science shows that the angular momentum and integral spin of the particle leave its centre at rest. Speaking of which, I might take some. What a jolly night that was. And now I'm proposing that this internal configuration amounts to solid rotation, or the essence of strong nuclear force. I think it's a non-linear mass-motion fusion with *spherical* spin. You wouldn't just have the outside bit turning. The whole formation, that's inside and out, should be dynamic. So it's not really at rest. Indeed, it should be the most intense form of motion going. Clearly though, every particle is not a spherical spinning force formation of strong nuclear force. But I think that the formation logistics carry through onto different levels. The grand scale cyclical one of our gravitational relativity just being the biggest one of them.

Physicists accept that four forces should interact for stellar existence. Why don't they apply it across the board? They can't. It's been impossible to distinguish between gravity and acceleration in a small region. And outside the atom gravity appears to overwhelm all other forces. I'm suggesting, reiteratingly sure, that there are links between gravity and the other three forces, within levels, that apply consistently.

Predicating the process on any level should be inherent motion, within spacetime. As such, we're not just dealing with the logistics of distance here. Even though, admittedly, distance make things difficult, I think we can still observe and measure a systematic interaction of four forces out there—or in here. Be it incredibly small or great, that is. With the spacetime concept we mightn't be getting the full reality of a system. We're not getting the nonlinear dimensions. This allows us to distinguish forces on every scale. So we can rationalise their interactions and transformations in various orders. We could draw correlations and extrapolate through them. Evidently, these four formations show correlations between micro and macro levels. Look at your planets and your orbital electrons. They each reflect a similar systematic set-up. But for us, they all define according to the linear plane of a massive spacetime curvature. And they do. Who says they don't. Everything's linear—to some extent. What is of interest is these things could also function according to their own systematic arrangement. That means the stellar system interrelates within itself, as well as through general relativity. Ditto the atomic system. Ditto your galaxy. And the big one of general relativity could have a deeper systematic raison d'être. It could have big time symmetrical implications. But it's these systematic processes that the linear dimensions of spacetime won't pick up. Theoretically we could be looking at a symmetrical pattern of four forces which defines eternal matter transformation. And that could form the basis of cosmic evolution.

A spiral galaxy, for instance, could describe either a contraction or expansion phase of transformations on a certain scale. The configuration expresses how spacetime and motion transforms matter and energy. It just happens to be a particularly big one. And it should be part of a particularly big force formation system. I don't see how a spiral forms completely of its own accord. It usually transpires out of spin and has wave affiliations. Anyway, the solar core empowering a spiral galaxy appears analogous of our theoretical spherical spin. You see; these formations show how the weak and strong nuclear forces might interact beyond the atomic system. As the scale increases they mightn't interact so much in tandem, for example. So I've already said that. Maybe it's a major point then. Anyway, recognising, and then systemising, these formations on different levels describes how the large scale and small scale aspects of physics might replicate to some degree, and interconnect. The cosmos could bifurcate.

Take the stellar life cycle. Within it could be formations that reflect the power of condensing spacetime with motion. A high mass star apparently evolves rapidly by developing the necessary pressures and densities to ignite thermonuclear reactions. Well, I'm wondering that it could be manifesting a strong nuclear spin. This is also a condensation of spacetime and motion; remember. The outpouring of energy from the thermonuclear reactions at the star's centre could resemble the atomic imbalance of a proton's spin. For me, that seems to trigger the charge system and radioactive release for stability. So I'm wondering that the stellar process could likewise reflect wave and spiral formations of angular momentum. Besides, that process apparently structures conditions by which the star can support the enormous weight of its outer layers. Is it therefore stabilising the balancing act in a system that supports planets in the orbitals? Is it functioning through a symmetrical process of force differentiation? And if all that's presumptious baloney, we've still got magnetic fields on a planetary level. They could amount to the waves of a *stellar system*—a macro version of your micro wave. ·

Maybe white dwarfs are massive expressions of a nuclear spherical spin. Well, ones without spiralling electrons in the atomic shell, that is. Perhaps pulsars of rotating neutron stars are emitting large scale manifestations of spiralling weak force radiation and electromagnetic waves. Further, that only the latter is discernible to us. Which could be why we don't get it that way. Is it so unbelievable?

Our finite thinking patterns delineated by the present laws of physics might only work out sequential elements. For that matter, we might not even be getting the sequence, just the elements. Black holes, big bangs, dark matter could be real only in so far as they present themselves as pieces to the greater Rubik's cube.

The nuclear mass inside the solar mass apparently counteracts gravity. When it stops, theories suggest a black hole of massive contraction occurs. In other words, the nuclear spherical spin is systematically framing gravitational rotation. When it ceases doing so, the orbital ring collapses in on itself. Maybe that's when the symmetry of the system kicks in. Another force interaction and formation would theoretically ensue. You could get massive weak force interaction occurring. That could be your accretion disc heralding in the black hole. And black holes, collectively, could herald in the amazing electromagnetic behaviour of quasars to realise a new gravitational predominance over galactic formations. Perhaps spiral galaxies signify the expansion or contraction stages of the greater cyclical set-up. That's supposedly the level of general relativity. Mind you, all of this is very general. We're skipping over puddles the size of oceans here with these scenarios. Where, for instance, is your micro correlation to a black hole in any of this? Perhaps it's not so definitive on the small scale. No one force seems to predominate there. You might never get an atomic hole because the other forces are continually firing away. All for one and one for all inside the atom. It seems more contingent. I mean the proton doesn't overwhelm the atom and eat everything in sight. Either way, the point is, a stellar system may not be the only arrangement to comprise of four force formation activity. And the stellar system could, well clearly it does, reside within the greater galactic one.

Now your black hole per se couldn't describe all of that. And you probably couldn't draw black hole correlations down the line. Your force formation, however, can. Or more to the point, your four force formation system can do that.

Even so, we've got oddities here, I know. But this issue: systems within systems, comes up in its own right further down the track. For now, I'm just gleaning that as we go up the scale in size these force formations don't seem to interact quite so intimately. They don't run in tandem so much. There seems to be more of a sequential interaction by them on the bigger scale. That might go on until finally; we're at the unilateral action of macro relativity. Further, all of these systems should impact upon each other. So whether a stellar black hole can instigate a galactic formation is indirectly probable, yet directly questionable. It would probably take loads of stellar ones. Anyway, I'll go into all of that later on. I'm calling it the bifurcating ladder. The main point here is that theoretically, within an infinite and eternal cosmos there are four invariant symmetrical structures, that replicate themselves within different levels, to form the transformation systems of the cosmos.

As with stars, galaxies, and other cosmic areas in the universe, there should be life spans for particles. According to theory here, they all develop, evolve and decay through transformation processes involving force formations. And that's right down to sub-atomic phenomena. In these next chapters I'm trying to describe how stuff inside the atom transforms through four different densities of spacetime and motion. These motion formations, evident on all scales within the universe, seem to reflect a code applicable to different scales.

In conclusion, here's another example; the relatively small volume of the galactic core. It apparently contains about 90% of the galactic mass. The system's spin? And 99% of the solar system's mass is in the sun. Likewise? These figures could correlate with the nuclear spherical spinning mass of the atomic nucleus. Remember; this formation has the greatest condensation of spacetime and motion. We have maximum motion within minimum spacetime. It realises a nonlinear mass-motion fusion. It's a symmetrical invariant component of a system. But let's go one step further. (This is possibly where I step over the cliff of credulity.) 90% of the mass of the universe supposedly consists of protons. And it would take 10-31 zero years, it seems, for half the protons in any given sample to decay.

Well, I'm wondering that here are some dimensions of a generally relative cycle. But most interestingly, it might also describe the links between systems. We could be gauging the domino effect of symmetrical bifurcation. Those protons could amount to a condensation of spacetime and motion showing the duration of the grand scale cycle via a slow release of angular momentum. Oh perhaps they are just being subjected to the boldness of an educated idiot. An idiot who's trying to make sense of eternity and infinity beyond her mental capacity and the observation and experiment of the day. And now I'm going away to wallow in self pity by the side of a trout stream.

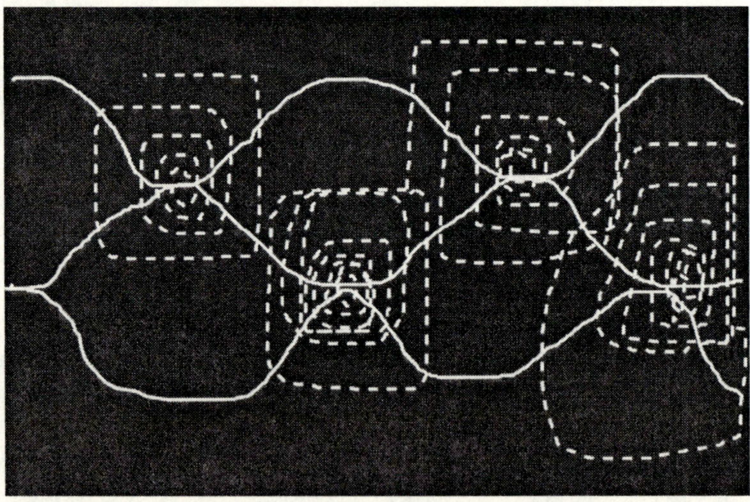

This is how our cycles of general relativity might look. As one expands, another contracts. The symmetrical balance therefore occurs amongst them. You don't just get one—there's no balance to that, no complete order. And only the eternal and infinite dimensions allow for this. Anyway, it can also mean; the predominance of one force, as happens on this level, say gravity, could co-ordinate with a different one next door. So the cycle here is not just what's within the blue enclosures. Nor do the red squiggles solely depict it. It's some of both. Plus that's a lateral drawing there.

Using a nonlinear dimension within an infinite and eternal framework you can reason all of that. Diagram-wise, without a hologram I can't really capture it. Imagine trying to work it out with numbers? I can't. Odds are on that one day another Euclid or Einstein will show up, though. Or will they? These days your up and coming genius probably has a struggle realising anything more than the dole, a habit, or a three bar riff. Not to mention the need to star in a head banging video with god awful lyrics. O.K. then, so what. It's a personal preference.

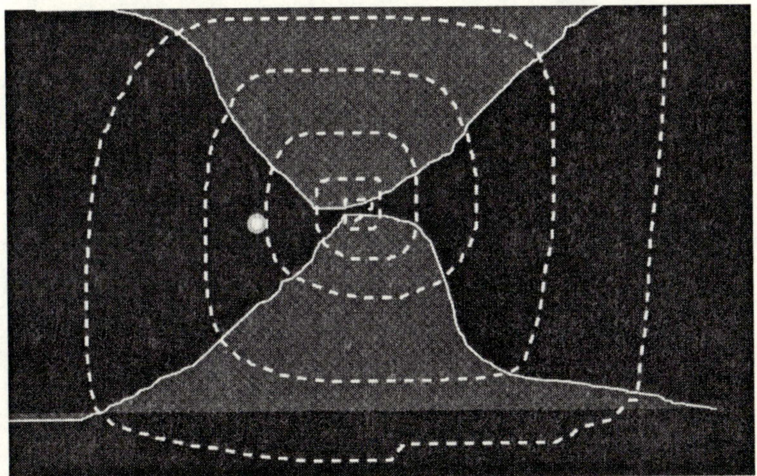

There's us looking out below the speed of light. Everything appears to be expanding away from us. And we can't find the dark matter to halt the expansion. I think we get the pink dots though. Theoretically, these are our cosmic rays of background radiation. They're supposedly weak force spiral remnants of nuclear debris, from a cyclical 'end-point'. However, piecing them together through our vantage point, and which is also subject to the linear framework, might only take us as far as the 'end-point' per se. Or, as in the diagram above, it might take us somewhere with no 'end' in sight. We mightn't get the continuity.

The formations of four forces seem to simplify complexities between different densities of spacetime and motion. Those formations describe cycles. I'm just hoping they also add up to power us in and out of this galaxy. I mean imagine getting back to find your local pub gone, along with everyone in it. Not that I go to them these days. . . .

. endnotes

nb 1 sorry for the repitition, but I must keep saying that—along with 'apparently' and 'it seems'—because nothing's certain about any of this

ch.4

conceptual development

. re-evaluation—keeping on track

> why is there something rather than nothing?
> because there is eternity and infinity

How big do you reckon the cosmos is? I've come to these theories and conclusions by accepting that it's infinite and eternal. And that's difficult enough to accept let alone grasp. Unless I keep re-stating the vastness of it nothing forthwith is going to make sense. Nothing might make sense anyway, but at least you'll know where I'm coming from. 'Infinity and eternity' forms the underlying premise for the entire book. So any order here that might unwittingly unfold—hitherto, comes from that premise. Chaos is going out the window with all other 'insurmountables', and symmetry's coming in. The thing is, wiithout an infinite and eternal playing field I don't see how symmetry and these theories would ever add up.

From the premise that the universe is eternal and infinite, one can address cosmological mysteries and paradoxes and come up with something. It's a double edged sword, though. Sure, such a platform provides the power to reach way beyond present thinking. Especially, that is, when you're not putting forward one lousy equation in support of anything. But on the other hand, your road is full of fields. There's no terra firma in any of these pages. So I could be coming up with stuff right off the track. What I'm getting at is grovel. If your mind's shut, we're wasting each other's time. You see; I'm not writing this to hone the irony of critics. And I'm not trying to enhance the skills of cryptic minds. Yet for all I know; you're a die hard realist. Trends point that way. So why read on? Enjoy your realities here in this solar system. These ideas probably wouldn't spring a leak with you let alone take flight. I guess I'm hoping for a thinking process. You know, if not .. then why not . . . ergo . . . something brilliant.

I could go through it all again with experts and libraries. But then nothing might come out in print for years and years, if at all. It's time consuming enough, at this late stage of the game, going through all the grammar with the computer check. Note; the short nifty sentences. And those eminencies aren't interested anyway in some unknown, unqualified, unpublished, unsubsidised nobody. And even if they were, which I know they're not, I'd have to dot every supposition with an equation. Or they'd make me feel so inadequate you'd end up with nothing, and they'd receive the regular gratuitous paper weight of a pedant. As it stands now, you're getting untarnished insight. If the ideas prove possible—well great eh. If not—then so what. You can bounce off them. Rise above the quagmire of post-modernism into a cosmological wonderland of infinite and endless possibilities.

It's an adventure anyway. Why should thinking only be subject to formal calculations or sacred laws of physics? This stuff's driven by wonder. That's that feeling excited by something strange—a mixture of surprise, curiosity, and sometimes awe. (Read the dictionary again.) Besides, because of adventures such as this, discoveries happen. You have an idea and then you go for it.

So despite everything I'm sticking with the initial premise; that the universe is eternal and infinite. And as the book progresses the conceptual framework for that should take on deeper meaning. To that end it's tempting to keep re-editing from the beginning—for the deeper meaning. I'm making the effort, however, to leave some things as they unfurl. Nevertheless, the idea is to emanate discovery rather than total stupidity, so I'd better start re-capping something lucid.

First then, our laws of physics only make limited predictions. Secondly; they don't seem to derive from an idea of infinity *and* eternity. Take general relativity. Underpinning it is the spacetime concept. Now I don't see how this can cover countless cycles within infinity. It's got that decisive beginning to it. So its predictions would always tie in with its framework. Even so, your concept based on eternity and infinity shouldn't invalidate general relativity. The deductions speak for themselves. I mean any idiot can see it's true. I just think a broader concept of the cosmos would show these truths to be conditional, rather than final. They could just be relevant to a phase in cosmic processes. It appears that general relativity places a limit on motion because of its conceptual basis, or its *framework for understanding*. To say it backwards then, the physical phenomena that the limited motion covers is just relevant to that.

The speed limit seems to have a dual effect. On both hands you never get beyond a certain point. Developing further laws of physics remains subject to observation and measurement. And your superluminal forms of measurement, particularly for observation, can be pointless through the existing laws of physics. It's twice as difficult because they work so well. Doesn't faze me though. Basing these theories on eternity and infinity provides the maximum perspective through which to analyse these laws. (If we can term my reading of them that.) For example, the speed c limit then translates into a conversion factor within transformation processes, rather than an insurmountable barrier.

Forms of measurement are subject to the laws of physics. And it's the contingency of this situation that the realist mightn't be able to address so far as the theorist can. If a law explains reality up to a certain level, the development of forms of measurement might only reach that level. So reality moves in, rather than on. The effect seems to highlight the limits of progress. And here's where I can step in. Although science relies heavily on observation and measurement, it's still subject to a conceptual framework of reality. The spacetime concept only recognises the reality within a set beginning and inexplicable ending. So we go from the big bang to the speed of light. And next thing we have Hubble's telescope producing images scientists accord with those limitations. It too is subject to the speed of light. We expand out from a big bang to an inexplicable ending. Just like champagne . . . to some.

How..ever, we have a law of physics that includes the speed of light squared. And it's some guiding light forthwith. The brilliance of this one equation through out these pages, for me, carries the same wattage as a tome of Riemannian geometry might to a mathematical giant. The questions flow accordingly. And when you link them to cosmic eternity and infinity, your answers become grist for the mill. You can end up reasoning transformation processes covering motion up to speed c squared. All of this translates through the concept of eternal cycles within infinity. It covers the beginnings and endings, and regulates the lot.

By applying the speed c limit as a conversion factor, waves and particles take on a different meaning. Their formations resemble unchanging symmetrical structures within transformation processes. If so, they could have extrapolation potential. We could gauge future processes through formations that the speed c barrier predicates. Your inexplicable ending can translate into another dimension. That means going full cycle through different combinations of energy and mass. $E=mc2$; you see, not mc. So the formation, be it wave or spherical, depends on a level of motion that is nonlinear. What this means precisely, forms a major section of analysis within the next chapters. For the present though, nonlinear motion theoretically has symmetrical invariance and is a *form* of the motion of space and time up to c squared. The wave for example, is nonlinear motion at speed c. The spiral goes beyond it. You wouldn't get these things with the present laws of physics. They forbid measuring anything physical beyond level c. At that point things either possess a position or motion, but never both. And I'm sorry about the gaps here. But this is just a re-cap re-evaluation; remember.

So our working concept for eternal infinity emerges. With motion in the spacetime concept we're over-riding beginnings and endings. Instead these limitations now signify change. Eternal cycles within infinity become the cosmic conceptual framework. The working part concerns its internal mechanics, which is where the equation fits in. I'm applying the formations of nonlinear conversions to the forces of nature. Here's our symmetrical invariance. What is symmetrical is nonlinear. What is nonlinear can also transcend speed c. And by page 100, if you're still reading the book you'll probably know what I mean. That's if you're not already nonlinear. Maybe your brain can handle whole chapters of this, but mine can't. I'm sure not writing it that way. Anyway, nonlinear and symmetrical are two key words, and there's no way round them. Later on there's bifurcating symmetry and home brew. And Sooty. Thank heavens for 2 golden green eyes in a sea of black velvet that go beyond syllables, completely.

Nonlinear motion needn't be chaos. I reckon it provides a structural involvement within a transformation process. Nonlinearity comes from motion within spacetime. You get your formation of spacetime and motion with it. Aligning that to a level of motion in space and time, or a force, provides an order. That makes for four force formations of spacetime and motion. In short; spacetime manifests nonlinearly into four forces of spacetime and motion. The one with the most space and time, and least motion, is gravity, with an orbital formation. And this is where the broader conceptual framework kicks in. Because its active range is immense, it's calculable to us on what appears as a linear framework. There's our general relativity below speed c that accords with the observable expansion within the universe. The spherical spinning formation of the strong nuclear force, however, has the shortest active range. It's more powerful in its microcosm, and clearly nonlinear. On that basis, the range of the force is also subject to its formation. And within all transformation processes, right up to the cyclical level of galaxies, these force formations are our symmetrical conversion agents, supposedly.

The ones that go beyond speed c are the spirals and spherical spin of the nuclear forces.

So there's the working framework for the next chapters. First up is a look at nonlinear motion via angular momentum. I'm trying to place theory with some of the facts. And who's denying it; my facts are surely questionable. My interpretation of them, that is. Four years' reading books on nuclear physics and cosmology in the bush, pre-Sooty and the power, makes me qualified to say nothing about nothing. No electricity, no supervision. I'm just picking up on Einstein's curiosity minus his genius, and gazing at the stars, sometimes sober. Great isn't it. And here we go with the role of electromagnetic charges next. Which then leads to questions on the neutron and how it fits into the scheme of things. Then I'm looking at the transformation of mass into motion, as distinct from mass into energy. Following that comes more theory and presumption. I want to tackle each force formation according to E=mc2. Well, at least draw the theory through its lime light in so far as your brainless wonder here can grasp it. These force formations are all subject to a level of motion. So it's just a matter of correlating the level to the equation. I guess as such, I'm also correlating the equation. Anyway, all of that should lead to further conceptual development.

Can symmetrical invariance apply as a conversion agent? Is it applicable within eternal transformation processes on greater levels?

You have to put it to the test, albeit if it's a hypothetical, incremental, mix and match one. And that's at your leisure, of course. I think the answer lives in that unknown territory separating general relativity and quantum theory. For these reasons I'll try to relate these laws of physics under the concept of infinity and eternity. Perhaps the concept will overcome their limitations. Well, what next? What did you expect? Audacious presumption is all par for the course. How are you going to get around the place superluminally if our 2 major laws of physics don't align? How's the planet going to fix the ozone layer while we're frigging around fighting each other? The logic stays the same.

So in short, on the basis of validating theory, I'm looking at how force formations manifest from 1 force throughout cyclical processes. The idea is to define a system that applies to different levels. That means you might find gravitational expansion on the same nonlinear scale as the nuclear forces. (Not the ones inside the atoms. I'm talking big time here.) Consequently, the concept goes deeper into the cosmos. Cycles take up the middle section of the book. I'm going into the expanding universe, a contracting one and cyclical connections. And that means getting into black holes, quasars and accretion discs. These strange nuts and bolts smell of superluminal possibilities. These things could signify systems that only an infinite and eternal cosmos can accommodate. They could signal the symmetrical changes in motion, space and time in which the speed c is but one.

ch.5

nonlinear implications

. *preamble*

Is physical reality simply spacetime, or a more diverse entity? General relativity and quantum theory both have reversible time frames that can produce a breakdown in time. They show processes and events where time can stand still, and where time and space tear apart. They express linear limits and their consequent problems. Beyond these boundaries is perhaps a greater universe. But we wouldn't see it under the present guidelines. And it's within the greater dimensions that I think we find our superluminal possibilities. I think we can only reach into it through motion within the spacetime concept. This means continual change. How can time stand still when it always has motion? Using it in the concept elucidates a dynamic nonlinear dimension. We've already gone through that. But for it to work successfully we should specifiy it according to the eternal and infinite framework.

For one thing, motion in the concept doesn't necessarily negate time reversibility. Taking the motion of an eternal dimension, however, at least provides the room to do so. You can fit cyclical forms of measurement into that framework. Through them we can measure continual processes. And everything can always advance forward then. There's no need to back-track or rebound off the spacetime limitations. Furthermore, cycles measure *completed* series of events following another set. Your cycle is sequential and systematic. Nor is it essentially linear. I think a spherical order incorporates the plane of polarisation, with cycles. Indeed cycles can amount to the mother of all nonlinear systems. Theoretically they're enfolding the nonlinear dimension on the grandest scale.

Still, scientists seem to think anything nonlinear is inherently unpredictable. Indeed, nonlinearity apparently signals chaos in a system. And the mathematics of it is a nightmare. We're talking differential equations, which express asymmetry yet both order and chaos. You must be joking. As a measuring device it can seem as confusing as the object of measurement, and yield equally confusing results. That's as we know it. Even I can recognise the complexities of it. So why not forget the maths, for the time being? Why not just give it a go schematically? First though, I think we need a clear idea and understanding of what it is to be 'nonlinear'. We need a simple version. And here's my a loop hole. Most dictionaries and science manuals don't even define it. So I'm defining it as I go along. And as we go it gets more specific and relevant to cosmic processes. Have you noticed? Call it conceptual development. Maybe the maths slot in later. I don't have the numerical tools, so I don't know about that anyway. How would I be, looking up the thesaurus for pi squared?

Even non-mathematrically this could prove confusing. So I'd better go over it again. For starters I don't reckon nonlinearity derives through complex mathematical formula. Well, what dimension does for that matter? You apply the maths to the dimension don't you? Or is that the pedantry of a mathematical midget surfacing again? Whatever. I'm defining this term according to a concept of eternity and infinity. Motion within the spacetime concept produces nonlinearity. It's that simple. It overcomes the linear plane of space and time. This opens the multi-dimensional window of reality. You're increasing the four dimensions of spacetime to a myriad. There's your continual change. You see; space can no longer begin with time. Instead, the concept now highlights continual processes. And it does this from the bottom line of reasoning. With spacetime and motion you're getting the bigger picture of causal processes transcending linear events.

Clearly spacetime becomes more complex. Even the slightest arbitrary influence in a nonlinear system renders countless possibilities. Yet it need not be hopelessly complex. Two simple regulations effect the possibilities: a spherical range for multi-dimensional

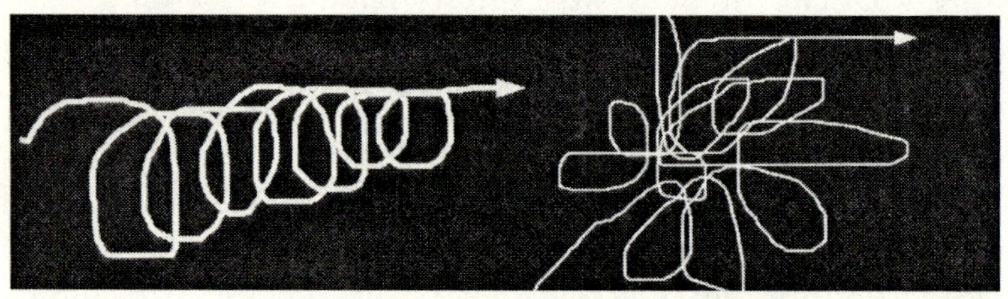

aspects; and an irreversible time frame simplifying the order. The former regulates the diversity of change, whilst the latter clarifies the unforeseeable. Evidently, the idea is to overcome the unforeseeable. This is why it works well with the macro measurement device of cycles. You can extrapolate full circle through it. Indeed, with nonlinear systems, change is a priori all the time. So we don't stop, ever. Equilibrium and entropy, for example, would reflect decisive changes rather than the story per se, or its end. This way we'd be addressing a chain of events.

The micro level isn't that simple, though. If indeed, the macro one is. Sure, it's still within a spherical framework of time irreversibility. It's just the protean nature of transformation processes that appears to stuff things up on this level. It makes everything hard to calculate. Spontaneity here seems magnifold. Energy, for instance, appears to transform *amid* the act of measurement. Our sub-atomic set-ups can display different behaviour in exactly the same circumstances. Look at the wave/particle duality of subatomic phenomena. Matter comes out as continuous with wave-like characteristics. And mathematics is a waste of time. Your numbers here can't distinguish between waves and particles. There's your nonlinear mathematical mountains for you; I wonder. And to cap it all is the uncertainty principle. This wonderful hum-drum allows for inaccurate observations. So, seeing is no longer believing. What a jolly scientific juggernaught that is.

Enter then, the nonlinear dimension of the space-time-*motion* concept. We can now consider wave-like properties of particles and the particle-like properties of waves as reflecting the forces of nature. Or should I say, their close knit association. It gives us four specific formations to play around with. And each one is subject to a level of motion, as well as space and time. Look at it this way; those formations could impose a symmetrical differentiation of spacetime and motion within transformation processes. The formation is a force. Moreover, they enact transformations. According to their interactions spacetime and motion translates into energy and matter. The thing is, on the micro level it's pretty well synergetic, for want of another term. n b 1 On the other hand, through the linear dimension of spacetime you mightn't be able to see into any synergism. You don't get all the aspects through which to find nuances. And after speed c you get nowhere anyway.

The nonlinear dimension, however, describes spiral and spherical formations that surpass speed c. These ones, theoretically, represent the nuclear forces of sub-atomic phenomena. Bear in mind though; the nonlinear dimension covers all four forces, on different levels. We can't just associate the nuclear mechanics of quantum theory with what's nonlinear. Then we're back on different planets again with our forces and laws of physics. Through the nonlinear dimension, we're looking at the symmetrical differentiation of *spacetime and motion* into *four forces*. We're not just reasoning the linearity of two of them through one particular framework. Nor are we working out bits and pieces of another on a different plane. Sure, the motion represented within the nuclear ones surpass speed c. But using this dimension we can get around that. I can theoretically correlate them to a force formation, two of which render mass and motion into a fusion beyond speed c.

Take the spirals of the weak nuclear force. Because they cover motion beyond speed c, they would not normally fit into the standard framework. Our measurement devices show fuzzy probabilities in between waves and particles. Sounds like a spiral doesn't it? Well, supposedly spiral spacetime spins around upon itself through intense motion. Let's say motion, to some extent overwhelms it. The energy and mass pertaining to one such force would be a mixture according to $E=mc2$. Which means it's neither a wave or a particle. The motion is fusing in with the mass. Maybe it's something like $ec+=mc+$. Either way we should go to Einstein. When in doubt—seek him out, or at least his equation. Placing it alongside these force formations you'd probably find different intricacies of transforming mass and energy. Of course, Einstein might not agree, not one iota.

Technically and theoretically, a spiral involves the release of angular momentum. It comes from the intensely contained motion of the strong nuclear force. Right. Well, how do you pin-point that? How do we transpose spiralling angular momentum with the uncertain calculations of quantum theory? And then come up with a practical explanation. Clearly I can't. But who could? Talk about technicalities anyway, and I'll only try and theorise myself out of it. But the thing remains, we're putting the dimension before the calculus rather than vice versa.

The interaction of force formations is a start. It's a basic platform from which to draw greater detail. If it's to be realistically effective, however, it should open technical doors. It's got to welcome in mathematical calculus and experiments at some stage. For now though, what we're coming up with applies across the board. The four force formations all fit inside the same nonlinear dimension of spacetime and motion. And for all these reasons, I'm hoping for an insight into rotational dynamics later. But that's about as promising as Sooty behaving for once. He's advanced from shy little kitten out of the wilds of Tasmania into the all time social butterfly of suburban townhouses. (Money doesn't grow on our trees, so we're here for a while.) You naughty little flirt. How am I ever going to explain the continuity between linear and nonlinear aspects more clearly with you demanding constant cuddly attention—from the neighbours? Oh, to go home. No more on the leash, Sooty. Loads of butterflies and grasshoppers. No jolly traffic. And you can come and go as you please. We can shoot the breeze through the valley to our hearts' content.

Oh and it might help to do the summary first, before wading through this chapter. The going can get pretty tough and tedious inside the atom. Nevertheless, there's things not in the summary that could spell it out for superluminal travel.

. *rotational dynamics*

Before getting into this, I want to clarify some points. Firstly, theory here suggests we have different levels of force formations. That means the general relativity of gravity could pertain to a level that sub-atomic phenomena might not. Well, at least not internally. But the four forces are the four forces no matter what level. They each have the same combination of spacetime and motion no matter on what scale you want to rationalise them. All I'm doing is systemising them for clarity. And ditto the same logic with linearity and nonlinearity, and likewise, rotational dynamics.

Next, these systems transcend each other. And that's where our confusion can arise. You've got your straight line of linearity as a small segment of a greater nonlinear framework, incredibly vast. But that same linear motion can still transcend sub-atomic phenomena. That's how I see the general relativity of gravity. According to the nonlinear dimensions of that vast framework, it would amount to an incredibly weak ratio within the curving order of things sub-atomically. It's impact, or dimension of curvature, could seem ephemerally inconsequential, to the point of irrelevance. You see; on the micro level, we're looking at intensely circular motion. We're looking at an atomic system that I'm suggesting has the scaled down *nonlinear* dimensions of the, as yet, undiscovered greater one of macro gravity.

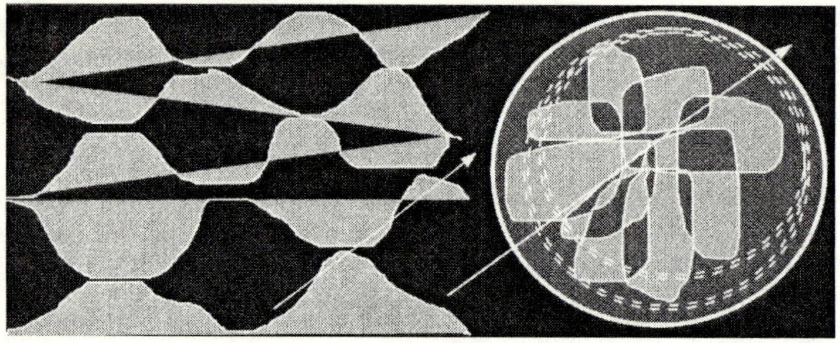

The light blue lines represent the 'straight' lines of general relativity gravity. On the left you can see how they would figure within a macro nonlinear framework. The scale has to be immense to get the nonlinearity. The stuff on the right is sub-atomic. Although it's way, way over-sized compared to the macro cyclical level, the same light blue line wouldn't figure sensibly as nonlinear within it. But externally, linear light blue should overwhelm the whole thing. Inside, you've got your pink, green and yellow squiggles happening in nonlinear disregard for the blue outside. In other words, the nonlinear scale of gravity renders it too weak to calculate on the sub-atomic level of nonlinearity. So the rotational dynamics of macro gravity could be a waste of time here when reasoning your photon wavelength, for instance. What it means for uniting general relativity with quantum theory could even be profound. I'm on a roll here. It might be theoretically amazing. Or it might be hogwash. And this reasoning, hogwash or otherwise, features later on.

So the nonlinear can exist within the linear. But get this. Clearly the linear is always within the nonlinear. You know; scientists don't seem to reason things this way. And no bloody wonder, a Cleesian comedy probably makes more scientific sense. Nevertheless, whether it's because the scale of nonlinearity on the gravitational level is so vast it defies observation for us, or because their reasoning derives from a concept of finitude are both probabilities. Quite likely the distinctions between the two (linear and nonlinear) define the disparities of quantum theory and general relativity. So at what level do we draw the line? When is the line no longer straight? Well, you can look at things systematically, which is what I'm doing. There's your nonlinear atomic set-up, for example. Then there's the stellar one, galactic one and grand-scale cyclical one. In other words, just take it that the cosmos is a nonlinear event. Forget linearity. And that's the simplest, if not the most irresponsible, uneducated thing I've said thus far. We're clearly forgoing the finer points of rotational dynamics and physics for theoretical lucidity here. Believe me I've tried though. At this editorial stage I'm confronting 10 pages of partly grasped stuff on angular momentum, linear motion, acceleration, torque, etc. You wouldn't want to read it. I sure don't feel like wading through it again. And if it wasn't for the speed of light I wouldn't bother. But it seems that the linear form of calculation is a part of that barrier.

Through the linear rationale, a force apparently equals the rate of change of momentum. Now wouldn't nonlinear motion produce more? The force formations illustrate this. Look at your orbital formation compared to your spherical spin. Which one packs the biggest punch? And for all I know the linear rationale mightn't calculate momentum via a formation. Well, not the ones I'm drawing. You'd stop at speed c with the linear rationale, whereas a formation carries on.

The nonlinear concept furthermore, shows motion is as much a physical quantity as is space and time. Spacetime and motion together, within one conceptual framework, recognise entire transformation processes. You don't just have E=mc2 but variations of e and m up to c squared. The possibilities should be manifold. You'd have permutations of mass, motion and their momentum beyond the linear dimensions. The concept depicts mass and motion through a formation, which is subject to its percentage of motion. So forces include mass—the mass of motion. Now don't ask me what that means for macro general relativity. I can only refer you to the pale blue lines of the previous diagram. The point here is the formation recognises the speed of light by its representation of mass as motion in space and time. And don't ask me about energy. You'd only be stretching the point of logic. Besides, I haven't worked it out.

These force formations cross over boundaries of what is linear and nonlinear. And their interactions might describe a transfer of angular momentum. On the micro level, electromagnetic waves interact with nuclear spirals and the orbital formations. Here's your nonlinear dimension of changing formations. That's full on rotational dynamics according to theory. But I'm further suggesting these formations are both active and inactive and can empower or stabilise each other. That's the systematic approach again. For instance, when the weak force interacts with the strong nuclear force it releases a spiral trajectory. This trajectory would be a product of the angular momentum from the spherical spinning formation. It's making it less inherently nonlinear, so to speak. And I hope you're still with me on this. Because now, I'm going to start flinging the finer points of physics around.

Take spin, for instance. It appears to fit in with the nonlinear concept. The spin factor seems to supersede the linear measurements of forces, or torque. At least that is, when you're calculating angular momentum on the sub-atomic level. Not that I understand torque. It just seems that rotation predominates over linear forms of measurement on this level. Through spin we could be looking at the nonlinear dynamics of the atomic *system*. And maybe we're not doing that. Are we calculating rotation from the meaning of atomic parts, as in systematic formations? Or do we search for the microcosm of mass, size and velocity that applies to the angular momentum of a linear scale? This system, ostensibly, might not figure by measuring mass with velocity, but through mass *in* velocity. That's vel-mass-ocity.

In the diagram below spin on the left is translating through four forms of rotation.

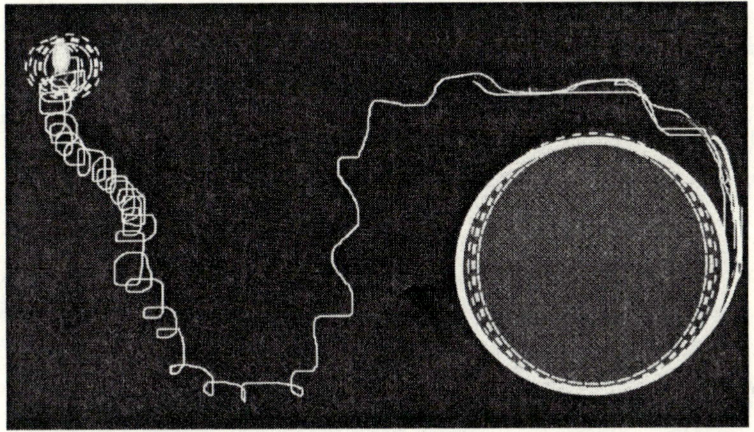

According to force formations, it's the motion inherent in space and time that forms rotation. And it applies to all scales. But on the sub-atomic level it should be more obvious. Theoretically, the spiral trajectory of the weak force carries some inherent spin from the strong force of the nucleus. Through the spiral formations the momentum of spin can empower wave formations. In other words, the spin can spiral out into a frequency. Finally, the wave frequency, with less motion and more spacetime can stabilise into the orbital formation of gravitational attraction. And that should still apply to any level, sub-atomic or galactic. So when motion is below speed c we'd always be able to calculate angular momentum through mass reacting to motion. There's your linear demarcation.

Now it seems we no longer see rotation as inherent within a system once we get outside elementary particles. What we have instead is the mass, momentum and velocity of general relativity. The linear demarcation covers the lot. And maybe the systems approach goes out the window.

The point is; force formations incorporate to varying degrees, all the factors of spin, up to speed c squared. And the only way I can distinguish the rotational dynamics of spin, beyond the orbital formation, is through the formation itself. But there's four in a system. And that includes some above and some below speed c. You're looking at them from the same viewpoint. It doesn't really say a lot though does it? Where's your centripetal and centrifugal forces in the formations of waves, spirals and spherical spin? They probably merge. How does torque figure within them? I can't say. All I'm saying is these things—force formations, can distinguish mass from motion, by the curvature of formation. They cover the processes of a system that can include the linear rationale. And I think that's necessary if we're to figure out mass travelling beyond speed c. Once the circle twists around through wave into spiral, then sphere, you're fusing your mass into motion. And that's when your space and time diminishes into the superluminal framework.

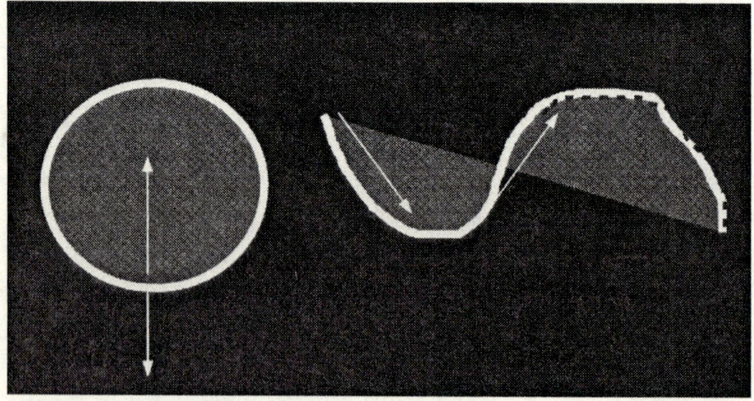

The blue arrows above signify centripetal and centrifugal forces. Incorporating them within the green wave formation should make it harder to work out angular momentum. Theory here is showing that it's only at speed c and beyond where these particular forces merge into the formation. And without the turning centripetal and centrifugal forces it's difficult to place momentum. Now interestingly, this is also the level where mass is invisible and indefinite to us. If we can't locate a centrifugal or centripetal force within the system then it could mean the system includes speeds at or beyond speed c. They're being overwhelmed and so are we in our calculations. So find the formation first, it's easier. Either way, all these factors could signify the defining barrier between linear and nonlinear calculations of angular momentum.

Through force formations however, mass to a degree, is acceleration. And here's something. Check out the missing dark matter. Where is it? *Is it part of the mass-motion that goes beyond light speed and our observational abilities?* Is it swirling around in the undiscovered nonlinear system that also comprises of a linear expansion? Is it part of the dimensions that are just too vast for us to contemplate? Is it in the spiral arms of a galaxy or the waves of quasar activity? Is it within the big bang of a stellar implosion? Is it a matter of next door's cyclical situation? We go out as they come in. Is it some of all of these things? Wherever it is, it's beside the point.

On the sub-atomic level things seem more unanimous. You're getting the spontaneity of team spirit in a nut-shell there. So it's easier to translate spin into orbital for the sake of the system. But that's still not the point. The percentage of mass in acceleration depends on the formation. And the formation is a *systematic event*. It all derives through a nonlinear dimension. It's just like the conservation of mass and energy. Spacetime and motion, in total, might never change. They should only translate into mass and energy via its symmetrical differentiation. So the rotational dynamics of our forces may not simply confine to mass in motion (that's the point). We may also need to consider mass-motion in spacetime—according to other variations divvied up through the atomic system.

Look, I clearly can't spell out the technicalities of this. Sorry, but I'm no nuclear physicist. And the only equation I can go by is our guiding light. But we know that when we hit speed c, mass increases beyond motion's linear capabilities to carry it. And this is also beyond our capabilities to reason it. We're going beyond the black and white of our clear cut laws of physics. Through standard reasoning it seems; before spin replaces momentum, and the turning forces go out of the equations, the system is linear. The only similarity between that and the nonlinear force formations is that this also occurs at light speed. What I'm therefore suggesting is spin could figure in calculations beyond and at speed c. Spin *signifies* the primacy of nonlinearity within a system. And we don't see it that way. Using these formations it's obvious. More to the point, using these four formations as a system, perhaps we could pick up the spin transition of mass. The formation covers motion, space and time transforming into mass and energy. The more spin, the more fusion between mass and motion. The less space and time, the more profuse are the curves of motion, and greater is the mass.

No doubt many particle physicists and anti matter theorists would disagree here. I beg to differ. From the waves of electromagnetism through to the spherical spinning formation of the strong nuclear force, mass and motion could combine in a way that embodies velocity up to speed c squared.

You know; it's difficult to open doors that allow for superluminal consideration. Putting differential non-linear equations aside, how can you calculate stuff beyond speed c anyway? Would our quantum number of spin get you there? I doubt it. It doesn't relate to the orbital for starters. The idea here is to calculate linear and nonlinear stuff on the same scale. Or, that is, calculate 'luminal', superluminal and 'lower-luminal' velocities through one set of equations. Again, theory points to the systems approach of forces. I think if you can grasp your forces through one dimension, then you're probably half way there. Understanding the percentage of mass within acceleration above and below that speed, might also help. I mean even if we assign a negative number for it. Maybe that's when mass is energy, according to E=mc2. But I guess what I'm really asking is; can any formation translate into a set of numbers?

Look, the linear logic is always within the nonlinear system. Who's denying it? It's just that our numbers don't account for all the forces of nature on one scale. The linear measurements of gravity, say, should always figure within other force formations to some degree. These formations are just derivatives of one force, anyway. So how do you find the linear set-up in the atom? How do you find the nonlinear set-up surrounding the line of general relativity?

I think I'll go for the linear within the atomic system first. For this reason I'm going into electric charges. Their positive and negative associations seem to represent a definitive direction, or line of motion. Besides, electrons apparently exist in different states called 'spinning up/down', etc. The up or down spin could tie them to the charge of a proton. I'm wondering that these electrons, and their charge function, can link the linear and the nonlinear aspects of a system. Maybe this direction of motion still calculates through the angular momentum of the linear framework. Perhaps the energy levels that we decipher could amount to the transference of angular momentum, or some rotational dynamics of spin, between proton and electron. Charge transfers could counteract proton spherical imbalance. And if we're calculating angular momentum according to the direction of motion of spin, as per electrons, then we may be dealing with the linear within the nonlinear already.

In the diagram below four spirals describe a transfer of angular momentum out of the spherical spinning nucleus. Each has a different direction of motion that represents the spin 'up', 'down', etc., states of the electron.

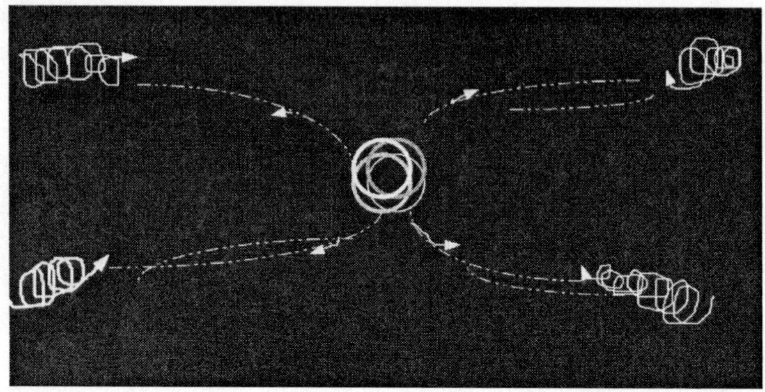

Clearly there's more to this then meets the eye. There's waves and orbitals and charge effects, for starters. The point I'm trying to make with this diagram, however, is that the superluminal tangle of nuclear spin can transpire and translate into motional directions.

. *charges*

There's another reason for getting into charges. Thus far, there's been no waves in the sub-atomic diagrams. If it's a micro nonlinear system of force formations they've got to be in there somewhere. I mean they have to figure systematically alongside orbitals, spirals and spherical spin. Well, I know they're there. Anyone knows there's waves with atoms. Your wave distances your oribtals. I just haven't figured out how they transpire through the hypo setup. So charges might lead us there. And henceforth we might get a decent diagram of them.

First though, some questions. Do charges express a line of direction that we can measure as linear? If so, does this line of direction readjust the nonlinear intensity caused by spherical spin? Is it an act of symmetry? Do charges maintain a micro system of force formations?

The charge could describe a linear dimension within the nonlinear. They reckon a system is nonlinear when its responses to charges are not just proportional to those charges. I take that to mean the system is greater than the linear format of charge directions. Now that's nothing new, I know. Clearly there's more to the atom than a positive and negative charge process. But how the charge fits into the scheme of things is the question. The charge could be kicking in the system's symmetry. But for all we know, charges—their raison d'être and origin, are as mysterious as the nonlinear system. If so, we're all in the same boat. Or are we? Maybe I'm paddling away in your Calypso Princess's lifeboat. Either way, I'm hoping charges will reveal more on the interactions of force formations. They might reveal further intricacies within the formations through linear dimensions. They could allude to the direction the formation takes when defining its shape. And they might also describe the systematic side of force formations, to wit, the micro one. Perhaps there's a symmetrical process between charges and the directions of a nonlinear system. Hopefully it's the same one force formations describe.

What then is charge? How does it arise and manifest? And what's its true nature? Why does it occur? And get this—its fundamental level is unknown. What a boon for theory. Now if that's not presenting a pre-schooler with play dough, then what is? Here we can go with the magical castles from the land of make believe, diehards. Your charge could figure in the structure of a force formation, say the wave formation, through an up and down motion between two things. If we can't create charge, its components might be innate within phenomena. Charges might arise as an effect of spacetime and motion. We could be looking at an inherent feature of force formations with this initial charge attraction. The charge could maintain and stabilise mass and motion within the 'infinities' of endless time, space and motion. They could be a symmetrical mechanism. You get a transference of some form of mass-motion between two bodies with them. That's acquiring charge. And here's the best bit; this is where one body has too much, and the other not enough, of the *same thing*. It's looking promising then. Symmetry kicking in.

Now theoretically, intrinsic motion *within* space and time, and not just its consequent mass, relates to charge. Charge should be part of the symmetrical differentiation of spacetime and motion into a system of four forces. So your motivating force of 2 charged bodies might just be a mechanism of all that. What happens, if for instance, motion becomes too intense within space and time? Remember in the nonlinear world you're always subject to change. It's dynamic. You could have a build up of stuff that throws things off kilter. Motion in a nut-shell might overwhelm space and time, with nowhere to go but into an increasingly dense concentration of mass. Wouldn't that make for some other form of spacetime with less motion, a stabilising prospect and a systematic component? Think of a pressure cooker.

At this nonlinear level the motion of spherical spinning formations, say of protons, supposedly calculates up to speed c squared. So there's plenty of motion to go around. Here electrons define through orbital formations as well as spirals. And I'm getting to the waves. All these formations could relate electrons to the concentrated mass and motion of a proton. All these interactions could reveal how mass stabilises and generates within the enormous power of intense motion, systematically.

Now here on the sub-atomic level, you might find an orchestration of directions of motion through charges. Look at the electrostatic force between positive and negative. That seems to equal the inward force of the electron in orbit, for starters. Plus these interactions and intricacies could mirror data from the linear dimension of the larger framework. Comparatively, you've still got the same vast spaces for gravity to figure as the macro level. And this is your nonlinear framework of a sub-atomic level. The atom mainly consists of 'empty space'. If so, the charge systems within atoms could have similar conditions of say, galactic space, where mass generates on the grand scale. I forewarned you. Here's the magical castles coming your way, out of play dough. And Sooty's doesn't give a rats. But that's after 8 hours of scrum and tackle, wrecking the blinds, wrecking my sleep, checking out the scene, shooting the suburban breeze. Go on sleep, little innocence.

So I'm wondering that the charge system is a mechanism for stabilising mass to the first degree of generating gravity. It seems to provide initial conditions for defining mass from motion. The negative, for instance, is subject to a larger degree of space. It therefore carries much less mass-motion fusion than the positive. All up, any mass of a charge system might be transferring angular momentum out of a highly nonlinear system. Via your charge directions, the electron's mass could correlate with an over-abundance of nonlinear momentum spiralling out of the proton sphere. The system's symmetry is kicking in.

Clearly our single negative electron formation is far different from the positive proton one. The friendly electron seemingly spreads around atomic space, distributing its charge through a cloud. And the proton is this tiny, anti-social, intense, massive formation just spinning around. It might take a lot of space to handle any proton release. Indeed, the space seems to afford the elongation of tightly bound angular momentum through into electron cloud. But can it tally up into the charge content of mass? We're reasoning *space*time and motion here; remember. I'm looking at energy and mass transforming through four formations of it. So either way, there should be more to this than just spirals and spin. Sure, theoretically, the charge mechanism has to deal with intense spherical spin. But you'd probably need loads of space to generate sufficiently powerful 'cross winds', say through orbital formations, against it as well.

Maybe the attraction between opposite charges develops into a restraining force against the intensity of spherical spin. Something centripetal could be going on. In other words, the charge attraction could activate linear dimensions. And some new force of spacetime and motion could develop. These lines of direction might change a spin into the frequency of a wave. Your orbital electron could be feeding off the wave transfer of spinning momentum. Here's our linear reasoning coming out. Here's fantasia in full flight. In any case, the process seems to define the orbital level and thus an electron's percentage of mass by level definition. Besides your orbital level, remember, is the below speed c formation through which you can figure mass in motion.

But the point here is that the orbital seems to be subject to the strength of attraction between negative and positive charges. We've got our system happening. Theoretically, the orbital comes out of a wave. You see; our charge here is structuring wave and indirectly orbital.

This diagram shows a lateral slice of a spherical spinning formation (way over-blown). The 'mass' of excessive nonlinear motion transforms through a linear direction into the rotation of orbital rings.

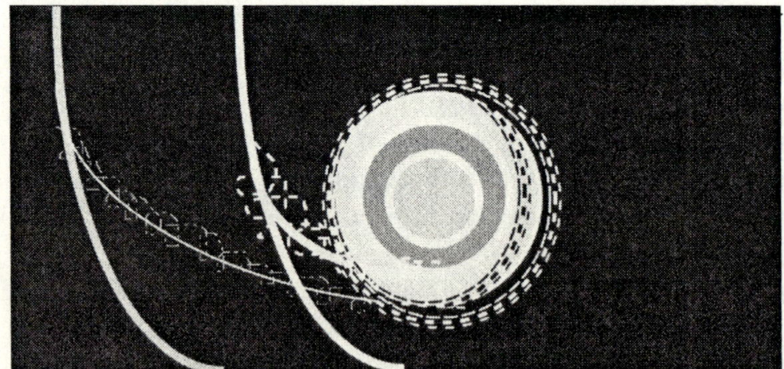

The positive formation of the proton appears incredibly nonlinear. It looks tiny, tight and introverted. Not that I've ever seen one, mind you. It reads that way though. Now contrast this with electron mobility. That motion formation could embody space. Rather than exhibit motional direction, electrons seem to absorb it. I think they're arresting looming infinities. Or the build up, that is, of an intensely nonlinear system. Technically, we're counteracting the positive charge of a proton with electron negative mobility. Theoretically however, we're diffusing a build-up of mass-motion fusion in the proton, through a dispersion of electron space and time. Indeed, electrons apparently have little mass. Should we condense their motion formation into the tight ball of spherical spin it would probably express bugger all velocity as well. They just eke out space and time instead. No wonder we don't detect them. All we get is an orbital 'cloud' and either position or velocity. I'm suggesting they're a systematic component of symmetry. One that is stabilising mass to the initial stage of registering a gravitational pull. In so doing, they're maintaining systematic equilibrium. And just using the linear calculation to locate them could obscure these nonlinear suppositions.

Quite likely, we're not contextualising our linear rationale. Well, O.K. then, there's no 'we' about this, mathematically speaking. So I'm assuming a mantle of nous with freelance authority. Whatever. But it looks like we're subverting the nonlinear to the linear. It looks like we're re-arranging E=mc2 into pared down m and e rather than systematic variations of both. And sure, using strength of charge to place the electron might show a linear calculation. But is that line just a general direction for the curves of four nonlinear formations? I look at the charge as expressing the 'winds of direction' within a system. One that goes beyond the linear boundaries of gravity and the present laws of physics.

The diagram below shows how the charge system might define orbital levels. Here the strength of attraction within the charge system describes the electron orbital. It's staggering excess angular momentum from proton spherical spin. And this makes for a distribution of mass, motion and energy in the atom. Presumably.

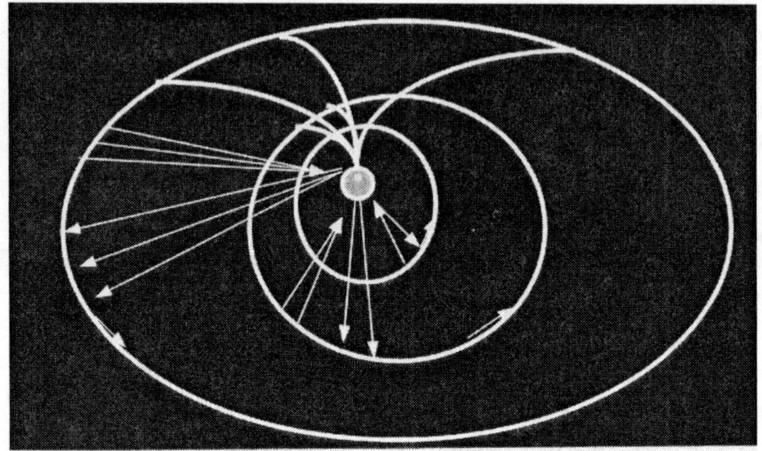

The thing is, these formations describe a nonlinear measure of space as well as motion, within a system. You can slot in the linear rationale here. I want to subvert the linear to the nonlinear. Theoretically then, our charge system is also securing a measure of space (and time) within continuous nonlinear motion. We're drawing a line in the sand with it. The charge system defines an orbit of specific angular momentum through which the electron can rotate. So the negative and positive situations are realising an inviolable space between. Without it, you don't get your symmetrical stability. The proton might spin out of control. Indeed, the strength of attraction between negative and positive determines how much space there is between orbital and spherical spin. Your power of attraction is the linear direction. And greater linear distance could mean more a greater transition of power. Look, zooming around the bigger orbital is an electron with more energy and mobility. Symmetry's kicking in through a transformation of space then.

Space provides room. Initially, that allows for converting intense nonlinear motion, of positive charge, into a hint of electron mass. As far as formations go, you're talking spiral elongation of spherical spin. Clearly there's more to it. The charge system doesn't just stretch out some spherical spin. It seems to define the space that allows that to occur. It defines the space to block any radial acceleration of electron energy. It presents, through specific orbital formations with electron mass, room for transverse action. Enter the cross-winds. Your converted energy can't spiral back into nonlinear spherical spin then. Enter the force of a wave formation. Here's something that can coordinate proton momentum between orbital levels. What I'm getting at is the charge system appears to relate to four force formations. So we could have some linear rationale defining all of them to an extent. It seems to spell out a system whereby nonlinear nuclear momentum transforms into the linear stability of gravitation pull.

Theoretically, interacting force formations control the spherical spinning formation of the strong nuclear force, through a trans*formation* of mass and motion. I see that as part of the stabilising process of a dynamic system. Without any such mechanism nonlinear intensity could overwhelm all else. You could get a black hole. How intense would your sphere of positive charge be then? Who knows, maybe theorists refer to a 'hole' as positive energy because there's none of the space that negative energy can provide, paradoxically. A hole doesn't have space? Anyway, if left unchecked, this positive energy should be the same as speeds approaching c. That's when you can get an infinite mass-motion exchange—if left unchecked. I think the electron interaction here contains and counteracts it. On a bigger level you'd probably be forestalling a black hole—come—quasar process.

Force formations should simplify all that reasoning. I'm hoping I haven't confused you off their merry-go-round though. Unfortunately, mixed with my technical ignorance they could bewilder rather than clarify. They could bewilder a statue of liberty into a grand canyon. Nevertheless, they should describe an order to spacetime and motion. Through formations of forces, spacetime and motion can empower or stabilise itself. Perhaps that's partly why we've got four. That's four shapes capable of expressing symmetrical invariance. We've got four different speed ranges with them.

These formations should cover, and organise, every possible direction and dimension that motion, space and time can take. They should include every little linear bit. And those electric and magnetic fields of charge input should be a big symmetrical part of the process. The wave force of it would effect decisive transformations of energy and mass.

Take the orbital formation. You can't take the wave yet, because I'm still working on it. We can define mass according to gravity through the orbital formation. This formation gives us the linear definition of phenomena. And the orbit might also present a direction traversing the velocity of proton spin. Theory however, suggests spin velocity is about as nonlinear as it comes. So finding a proton direction with a right angle to the electron motion is pushing it. But we've got our charge direction. There's your line in the system. If we can trace it out according to the electron situation, we might have a linear definition of a gravitational force *within* a nonlinear system. And that's on the sub-atomic level. That's reaching right into the heartland of quantum theory. It means we're defining gravity alongside the high velocities of nuclear power. That's using force formations. Technically anyway, positive and negative pressures of charge together could produce gravitational effects. That's if they are a mechanism that manifests mass. I know it sounds pedantic. But here is a line of logic. Where you can locate mass you should be able to locate the gravitational force, even on a micro scale.

Now we're touching on deeper issues. My grumbling stomache, for one. And then how does $E=mc2$ fit into this? I want to spell out force formations with the mass/energy transformation processes according to that equation. Precisely how does the formation develop through charges? What does that mean for the speed range of each formation? How can you go beyond speed c without your mass becoming energy? First I'm having some egg flower soup and barbequed pork in plum sauce. Then I think we need to address the directions of charges, and how they form. And that means getting into electromagnetism.

. electromagnetism

So in we go, deeper into the convoluted world of nonlinearity. Maths could be a blessed relief. Thus far, I've outlined linear directions of motion in the nonlinearity of the atomic system through force formations forming and functioning in unison. Charge systems have a lot to answer for. I'd like to leave them out, but they're supposedly proving points. Besides, they might present a control lever for manipulating forces. At this stage they're coming over as an avenue for reasoning mass and gravity within atomic systems. You can decipher the formation of forces through them. But we still can't say theoretically for sure yet, that the atomic system comprises of four force formations. For this reason I'm going to throw electromagnetism at them as well. I want to work out waves through the charge system. A wave length separates the orbitals. But how does it happen? How does the moving charge define the wave formation? How does that figure with light transfer at speed c?

So the big ask continues. But nothing's insurmountable. Not now Sooty's back. What a leap off the balcony that was. What a three hour wrench of the heart strings, while cheeky flirty forged the traffic to check out the neighbours across the road. You can take the boy out of the bush, but you can't take the wonderlust out of my Sooty. Hey ho then for the open road, little one. Only 6 weeks to go.

Well, let's clarify this convolution. With charges I'm looking at motion directions within a force system. We're supposedly getting the linear dimension in the nonlinear context. Certain directions can show, for example, that angular momentum mightn't derive through the orbital plane of linearity. Instead it could link into intense nonlinear motion. It can come out of strong force spherical spin. That's your positive charge set-up. And if so, angular momentum should be able to translate the nonlinear into the linear. Or more precisely, it might be able to carry forward linear calculations. Furthermore, these linear inroads should fashion our nonlinear force formations which here, on the sub-atomic level, seemingly perform systematically in practicable unison.

Now I know focussing on the linear inside the nonlinear dimension might seem atomically absurd. But how else are you going to make sense of the nonlinear? Look, no tellurian technician has thus far aligned our laws of physics through any numerical expression. Those charge wizards could be addressing superluminal nuclear systems. Too bad I'm ignorant on the technicalities. What I'm grasping though, is the linear line of reasoning seems to highlight a transverse function of charges. This is one that stabilises and transforms nonlinear motion. And here on the sub-atomic level we're up against a gamut of it. We're also looking at mass and energy transforming in a concurrent format with these forces forming. On that basis, charge systems might hold clues for manoeuvring nuclear power systematically. They might reveal how you can contain mass beyond speed c. And there's loads more questions to that end. Buggered if I'm posing them. I'm questioned out for now. Just take it that they'd ultimately concern the manifestation of power, and the stability of mass, beyond speed c.

Our wave format is the first step across that barrier. Now we're merging the linear into the nonlinear. Theoretically, the intense rotation of a spherical formation initiates the charge system, and the formation of waves. Clearly spinning motion doesn't just produce magnetism. You also have a direction of spin here. Your magnetic field ushers in winds of direction. There's your range within which *lines* of direction are happening. It's not a nonlinear complex of introverting motion, in other words. And one field could be transverse to another. With the electric field we should end up with wave formations. Together they should form the electromagnetic force. In short; by segregating some intense nonlinear motion of spherical spin through two fields of space, I'm getting a bilateral formation of waves. And that's not totally linear. But neither is it full blown nonlinearity. With the wave we're getting more space and time, and less motion, than the nonlinearity of spirals and spin.

In some sense, I'm unravelling solid nonlinear motion by drawing it out through linear directions of space. In so doing, mass and energy should transform and flow. You can get light flowing along the waves in packets of photons, for instance. By imposing space barriers, as in fields, within nonlinear systems, the charge system can use directions of motion to stabilise mass-motion intensity. It could be a spacetime lever for nonlinear velocity.

To figure that one out, I think we should dissect the wave some more. Perhaps then, an effect of the magnetic side of things is straightening out excess motion. So the field surrounding spherical spin wouldn't change spin velocity per se. Rather, its nonlinear build-up would be animating space around it. Again, the symmetry's kicking in through space. Without the surrounding field, strong force spin might manifest some other change. It might even black-hole itself or explode. Whatever. This magnetic stage of the charge system could be the first step towards a level of velocity. The more decisive lever seems to come from the electron orbitals. Here we're stabilising excess spin through the cross winds of linear formations. I think these directions of motion—the cross winds, establish different levels of motion through a transformation of mass and energy. If so, we're looking at concurrent energy and mass levels rather than just a nonlinear fusion of mass-motion in space and time. Orbiting electrons never have the same level. The greater the radius of curvature, the higher is the speed. The longer is the orbital line of linearity. The more space it takes up. And the electron in the farthest orbital has the greatest velocity.

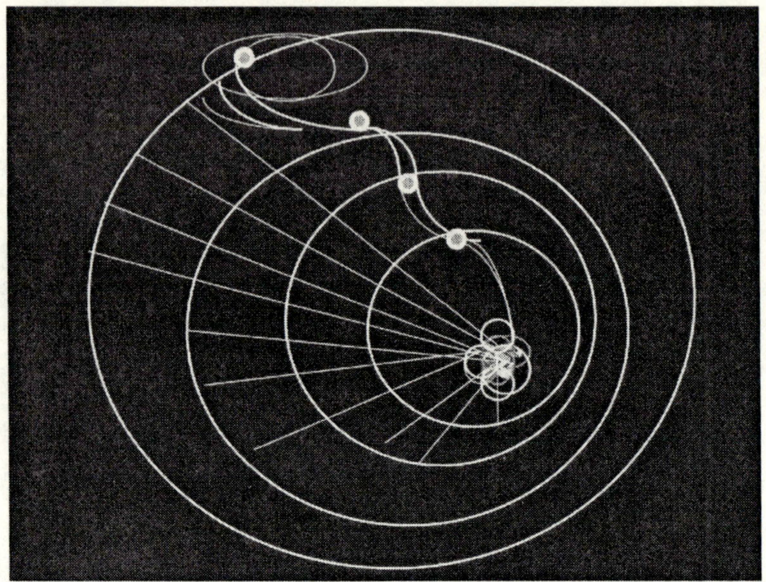

In the above diagram a wave formation results from the imposition of an electron's magnetic force. But each electron, and its orbital, is subject to its force of attraction with the central nonlinear motion.

Broadly speaking, here is a sub-atomic system showing the radiation of intense nuclear spherical spin. However, I think a particular stage of intensity involving a force within spacetime and motion, on any scale, will activate the formation and response of another force. It's just that on the micro level this happens all the time. Your four forces are forming and firing away together with the charges fine tuning their balance. Your electron obitals automatically adjust where energy is too intense, according to the spherical spin of the proton. The charge system just co-ordinates the process through linear dimensions. It levers the velocity of spin according to transverse formations. Evidently, these are inherent features of spacetime and motion. There seems to be a structure and a mechanism to it that perpetuates cycles of evolution. The differentiation of one force into four could be one of these inherent mechanisms. And I reckon their sub-atomic production fits into the greater scheme of things.

What would happen if a charge system didn't interact with the proton? Here's the potential for one force to manifest out of control, overwhelming all else. Without the symmetrical differentiation of spacetime and motion into four force formations, you don't get the show on the road. Experiments regularly reveal the potential for rising infinities. The differentiation of spacetime and motion through force formations however, seems to counteract any rampant augmentation of one specific force. It's the logic of systems. It's the raison d'être of charges. They produce: space formations, cross winds, and motional directions, amounting to different force formations. And these formations allow for a variegated interaction with any spiralling infinity of one force.

Indeed, our facet of infinity could stem from a one force situation. What about those spiralling infinities that manifest a uniform nature? Take the infinity, for instance, of overwhelming gravity. What about the unbelievable intensity of a 'big bang'? Does it arise from the infinity of nothing? It defines infinity through a beginning. Now just what your charge system amounts to on that large scale set-up I don't know. I've got a feeling though; the quasar expresses massive waves. And it seems that as the force formation scale increases, they don't function quite so concurrently. So there probably is no direct analogy anyway. Even so, these four forces theoretically always act systematically for a symmetrical reason. Theoretically, you wouldn't get a black hole overwhelming the cosmos. You'd still have your magnetic fields and electric fields adjusting scenarios in spacetime and motion.

I even think that a charge system presents a capacity for eternity. Look I'm full of it. As the instrument of force differentiation, it controls the infinite aspect. And theoretically, it seems that it is the eternal dimension that controls the infinite dimension. (Although it could be more contingent.) Anyway, through four different force formations entire transformation processes can take place. They can *continually* take place if forces keep differentiating like this—if the charge system does its thing. Correlate it and there's your cycle's of evolution happening. But back to the point of waves.

Waves—what was their point? I think I'm past it whatever it was. I wouldn't mind checking out some wet ones right now. But it's too far to walk and I'll lose the convoluted plot. And I'm sure not sitting in the sand with a portable computer trying to figure this stuff out. So where's Sooty? He might want to check out the lizard situation. Lizards or surf then, the forgotten point can wait.

With waves we also have the wave-particle duality. I think that suggests the meeting of linear and nonlinear events. Experiments quantify the energy transferring between orbitals as particles with zero rest mass. These particles, or photons, are apparently the transmission agency of electromagnetism. Well, why is there a particle and a wave together? According to my theory, waves express mass and motion at speed c. Particles express mass moving in an orbit below speed c. Instead, we have a confusion of both at the speed limits of the laws of physics.

Evidently, on the sub-atomic level it's difficult to distinguish mass and motion for any appreciable amount of time. Now that's not just your spacetime continuum in a nut-shell. The intense nonlinearity of a sub-atomic system seems to render its processes fuzzy. Supposedly, of the four functioning in unison here, only one of them realises a clear mass situation anyway. In other words, this is unlike the macro set-up. That's where one force seems to predominate over the others. Here they're all on equal footing. And it should render the nonlinearity of mass-motion fusion pretty prevalent.

Results of experiments show the electron doesn't move in a clearly defined orbit formation. It gets around in a cloudy one. Even the wave-length separating these indistinctions is questionable. Questionable, in so far as it has a wave-particle dual nature. Force formations however, total four. That's on a velocity scale covering energy and mass transformation up to speed c squared. Now that's some nonlinear scope for describing transformations of motion and mass. And despite the sub-atomic social scene, these formations shouldn't readily transform into each other from opposite ends of their velocity scale. The charge system wouldn't just manifest orbital out of spin without spiral and wave input in between. Thus the electron and photon would not simply transpire out of nuclear spherical spin. Somewhere in the process, waves would modify spirals. And orbitals would influence waves. On this basis, the electron orbital could scramble the wave. The electron's mass is no clear cut event here. Your linear diffusions of nonlinearity is no predominant act. And the consequent energy and mass transformations should follow suit. It means in short, that the electron is not simply mass *in* motion, under the guise of micro gravity. You don't get it under the linear dimension like you can outside the atom.

Besides, this wave formation of spacetime and motion only accounts for speed c. In one such force formation, mass only calculates with a zero rest value. It's nonlinear fusion with motion is still evident. I think for the energy to differentiate from mass, you'd need to involve the linear lines of orbital space ratios. On the other hand, if we upped the wave frequency into a spiral, mass would be minus zero. We'd be calculating the nonlinear fusion for it, again, more so.

Perhaps the electron is always subject to a nonlinear combination of mass *and* motion. Our glimpses, through orbital clarity, of mass in motion would be subject to a very dynamic system. Look at the emission and absorption process of photons. Well, you can't really.

The photo-electric effect predicts instantaneous energy transmission. How is this possible? One way is through spiral formations. And theory thus far, has spirals happening inside the atom along with the jolly rest. There could always be a superluminal spiral on hand to boost your wave transmission.

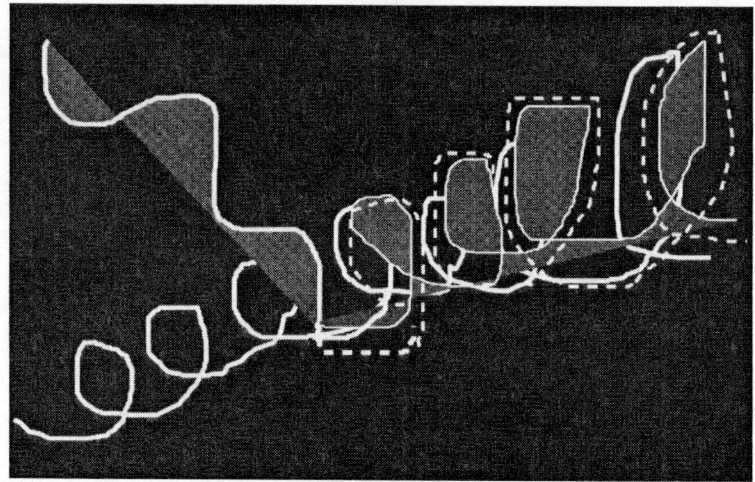

Hypothetically, that is.

Theoretically, the emission and absorption process of electromagnetism involves the transformation of intense nonlinear angular momentum. Electrons and photons wouldn't just be agents of this process. These little numbers happen along the way. I think they're our initial mass and energy transformations as forces differentiate and form. And they're transpiring from a *system* here. And pardon the reiteration, but it's one involving velocity levels up to speed c squared. So your spirals and spherical spin could figure just as prominently as your waves and orbitals. Both of these forms could underwrite electromagnetic formation. In other words, inside the atom, we should have a heavy influence of superluminal forces with our photons and electrons. Here your superluminal business doesn't just erupt over a predominant period. It's figuring and distributing all the sub-atomic time alongside atomic gravity, et.al.

Sure, our energy and mass transformations seem to derive by a diffusion of the nonlinear through the linear. But there's no sub-version of the nonlinear dimension to it, per se. Besides, the charge system, whilst making linear inroads through spherical spin, doesn't function without nonlinearity.

The spiral is what I'm getting at. It's always there. More often innate and inherent than not, maybe. But it keeps showing up, and not just theoretically. As a formation of weak nuclear force, it's closest on the velocity scale to c squared. It's only active within the most intense nonlinearity of the proton. And when it all boils down, spiral is just a bit of spin. But this bit seems to have a lot of potential. As spherical spin travels faster than any other formation, a recoil through spiral potential into a spherical formation could save time through motion. It could up the pace of the photon process.

Even so, a photon doesn't travel beyond the wavelength velocity. It's just a quantum effect of it. I'm wondering that it's a recoil of the wave spacetime and motion combination. Let's say it's a trans-formation element of this force formation. One that is not so much subsequent to, or even consequent of it. Yes, I know I'm mincing words here. But get the nuances. Anyway, the sub-atomic quan-tum of photon could still be akin to the symmetry of the wave. So any different velocities—according to the photon, that transpire through spirals into spherical spin, or even from orbital interfer-ence, would come and go within the wavelength. Your photon always appears with zero rest mass, moving at speed c. Here again though, is just another hypo scenario using the logic of theory more than hard facts. And below are some of those twists and turns.

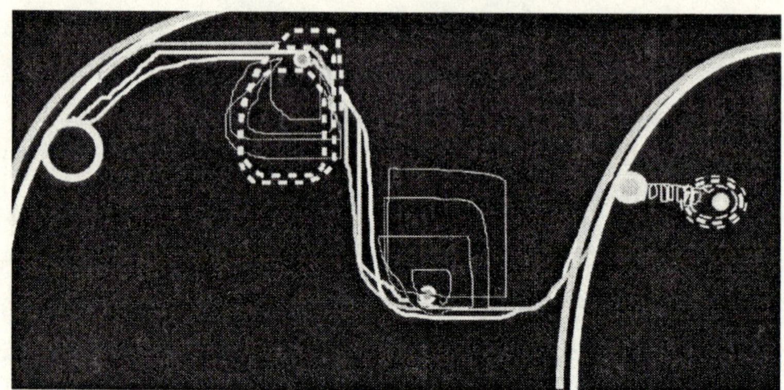

That's how some wave might spiral into a spherical photon. Your pink and yellow photons stay subject to the wave formation and its motion. Likewise, the yellow and green electrons remain subject to the green orbital levels. And everything black is nonlinear. It's a nonlinear system in a nonlinear dimension. It's a nonlinear world. That's the theory.

It's amazing just how creative these twists and turns become in your imagination. Who's to say reality is otherwise, though? Nonlinear equations? The uncertainty principle? Perhaps the wave motion can act on an inherent spiral formation. And maybe spirals can act on the spherical formation. If so, they could influence both the wave and particle-like aspects of electron energy. If energy comes out of a spherical spinning formation in spirals, the wave can absorb it. The linear direction of charge could be its guide. Are you with me? I've got spirals coming out of spherical spin in a linear line of charge format. I suppose the uncertainty principle is looking pretty damn simple by now. Pretty damn right, even. Well, there's no back-tracking now. I might as well go the full Monty with this.

Our electron could be transferring energy kinetically through a 'spiralised' photon. That's while it remains in orbit. It's probably the most efficient way of transferring energy through charge systems. If so, then it could work. Why waste spacetime and motion?

Symmetry doesn't do that. Look, when your photon strikes an atom, boosting an electron into a higher orbital, it seems to do so instantaneously. You don't see it traversing the intervening space. So I'm wondering if spirals and a charge system, yes, them of the convoluted discourse, can also generate superluminal motion for transporting mass. Or do they just sort out the nonlinear fusion for transforming it?

What about the role of the neutron? And perhaps we should spell out radioactivity according to this theory. And perhaps the footy's on and there's a cold beer in the fridge. I know there's a cold beer in the fridge. Besides, it's Saturday. I'd like to put neutrons and radiation on a boat to nowhere. But we've yet to arrive at enough insight for getting off the ground superluminally, spirals or no spirals. And they're laughing next door. That's right, drinking and laughing on a sunny afternoon in a suburban sprawl. What next.

Before going on, I'm going to flash-back and recap some. I'm not ready for any more sub-atomic subjects yet. So far, spirals show up as a definable mechanism of intense velocity. And they seem capable of generating superluminal activity, within the electromagnetic process. Their format embodies a nonlinear fusion of mass and motion, albeit less dense than the spherical spin. Should we unravel them further? I've got a feeling we're up for spirals with neutrons and spirals with radioactivity later. For now though, we're up against a juncture of the laws of physics. They don't just refute superluminal motion. Through them, the wave-particle mystery is proven practice. On that basis, I don't think you can't just refer to the wave-particle duality as spiral interaction, and take it from there. It's tempting though.

To date, looking at charge systems, and the electromagnetic process, through force formations reveals a method of sub-atomic transformation. At least to me it is. You might think it reveals bugger all besides a tangled mess. Could that form the mechanical basics of spacetime and motion? One that relates to superluminal activity. From the intense nonlinear power of the positive proton charge, through to the stability of mass, definitive directions of motion look like a causal mechanism. What I'm saying is there's a method to the madness.

Theoretically, the bottom line is four formations are manifestations of one force. On face value, velocity seems to distinguish between them. The charge system and the electromagnetic process however, reveal a more complex situation. They show that velocity is not the only factor defining each formation. Charge systems of electromagnetism describe fields of space and directions of motion as determining factors of a force. No doubt, time is the third one, but more about that later. The thing is, how a force figures through a formation should open up the equation $F=ma$. We're dissecting mass here through the systematic terms of spacetime and directions of motion.

Another point that seems to have arisen is the sequence of these forces. Their formations don't directly interact unless they're adjacent on the velocity scale. A gravitational orbit would not, for example, spring out of the nuclear spherical spinning formation. Even so, all four force formations appear to be either actively or inherently present. The charge system drawing into the electromagnetic process shows that. Maybe there's something in this then. A capability to produce intense velocities could still co-exist with the stability of mass in a gravitational force field.

Descending a velocity scale, using spirals, waves, and orbital formations, we arrive at the stability of mass in motion. And that's all out of spherical spin. It looks like the linear directions of a charge system orchestrate the lot. Through orbital corridors of electron currents, the electromagnetic process transpires. In so doing, it reduces some momentum of nuclear spherical spin to the speed of light. You also get the stability of mass in motion. Further, an initial charge component seems proportional to the resulting mass. And all of these forces co-exist within the system. There again is our gem of theoretical wisdom. The intense velocities of nuclear spherical spin are not in total isolation from the mass of orbital rotation. So speed c squared, can supposedly happen along side mass in orbit. And it's not worth a pinch of salt to us. Technically, these forces don't figure beyond the c.

. charge formations with radioactivity

And now we're looking at radioactivity through the force formations. That'll be a theoretical exercise and a half. The dimensions though, are still on par with the dimensions to date. There's no getting away from them. We're nonlinear all the way. The whole book's nonlinear. And if it isn't creating one, theory should be clearing a pathway through its dimensional thicket. You see; the structure of force formations should apply to everything, basically. Despite how I'm presenting them, they should be simple mechanisms of the cosmos. You can't get much simpler than a circle, wave, spiral and sphere. It's just their consequences can become as complex as complex. I'm just hoping I don't end up in a rabbit warren of convolutions. The idea is to keep applying them to the facets of physics that we understand. Or should that read—to the smidgens of it that I've cottoned onto? Anyway the idea is to test the theory. And the questions along the way are for this and the superluminal reason. I don't see how you can do that from within the confines of the laws of physics. You couldn't pose the same questions—let alone draw answers.

Through radiation, a mechanical nature of force formations might become apparent. It's what I'm aiming for—something tangible that a rocket scientist can rev up. As yet, all I'm getting is an active and inherent side to the interactions of these formations. And that's still unclear. So we have radioactivity. It's just another way of looking at electromagnetism to me. The incisive differences however, are probably profoundly important. Nevertheless, what I'm grasping is this.

Through radioaction formations seem to re-apply back down their velocity scale. For this reason, I'm thinking it might show how forces function systematically *within* each other's limitations. Theoretically they obviously do. You've got your atom happening within your stellar system. And you've got your stellar system happening inside your galactic one. How do they do it? Could we shove the mass of a gravitational orbital alongside the intense velocities of nuclear spherical spin? Can established mass travel beyond the wave formations of the speed c without becoming a nonlinear fusion with motion? Could Sooty make it to Delta Pavoris and back without becoming promite?

In the process of radioactivity we're revisiting the contributing formations of electromagnetism. They're the ones that structure the wave. I think it acts upon their manifestations. That means radioactivity could derive through unravelling the product of force formations' interactions. It's tapping into the force *system* and breaking it up, somewhat. Not like a virus. Even if this is taking on the inverted mantra of the previous convolutions. That's radioactivity? Read on, it might get worse. The released result, or radiation, varies. One form is more agile than another. Now is this due to the interplay of force formations? I mean is it because they're inherent or active, that decides the outcome? The more you chop and change your piece of spacetime and motion in and out of energy and mass, the more nimble it becomes? How do you work that out?

How does the wave function to realise different levels of energy? I know you've got your frequency and wavelength. But we're in force formation land here, remember. A greater frequency could mean more spiral involvement. You could have an inherent element determining the outcome.

The basis for all radioactive interactions is the wave formation—I'll get into the alpha and beta business later. The wave's confusing enough for now. From what I can gather radioactivity won't happen without the electromagnetic animation of space into the wave formation. However, the wave doesn't directly act on spherical spin. The wave requires a spiral transition before it actively releases energy from that formation, as per theory thus far. The question is then; to what extent are spirals and waves inherent or active?

How do we decide what percentage of a force formation is inherent and what percentage is active? Where does the wave start and the orbit end? When do I knock offf? They're all in there and I want to find out because later down the track you might be able to strum up superluminal power using inherent and active spiral possibilities. Theoretically our electromagnetic process of waves and radioactivity stem from a charge system that itself arises from the highest speed factors of nuclear spherical spin. We've already done that. The linear set-up of the wave is a systematic event that sprouts through four formations.

In a wave formation, the imposition of the gravitational force could render wave strength practically negligible in some areas. The weaker formation supposedly acts on the stronger one, making it inherent to an extent. Besides, experiments show that greater gravity affects the electromagnetic wave. Once outside the atom it surely would. But inside the atom, gravitational affects might not amount to velocity—at least not directly. Theoretically the wave is speed c on any bifurcating level you want to reason it, anyway. The wave has other attributes. These could differ according to the rate of systematic force interaction—that being subject to the scale of the system. And sub-atomically your wave would have more opportunities to interact with spiral formations than on a bigger scale. And that also applies to a gravitational orbital.

Now, I'm thinking the orbital signifies mass in motion. And waves mean a combination of mass and motion. The wave formation confuses them however. The wave can translate mass into motion at speed c. So what about these different attributes? Where one form of radiation can interact with matter, another can't. I'm putting it down to the rotational dynamics of a sub-atomic system. That's a mixture of inherent and active force formations, with all manner of directions. A spiral has some frequency and some spin and goes faster than an orbital. And the orbital only has a 'linear' directional movement that generates a one way attractive force of gravity. These directions of motion could empower waves in different ways re their agility—re their frequency. The one with more inherent spin potential a la spiral interaction could have a tighter frequency and penetrate mass more, for instance. Your wave frequency then might signify just how socially active it's been—inside the atom.

But let me look at this through the foggy glasses of particle physics. Sorry, that should read; let me look at this through my foggy understanding of particle physics. Take your 'half lives'. Would the time taken for half of any mass of an isotope to decay completely, amount to the unraveling of a nonlinear formation through its various permutations? I don't know. I'm just setting up the corollary. It could range from fractions of seconds to billions of years. And then you've got your motion/space factors of time dilation. Perhaps it's easier just to address your protons and *how* they equate to your neutrons. Either way, all these factors should relate to inherent or active aspects of force formations' interactions. The time it takes for a radioactive source to decay into half its original value, should correlate with nonlinear spherical spin and E=mc2. And it should say something re the agility of the pursuant wave. And I did work it out, you know. Roughly, that is, very roughly. Each force theoretically tallies up with a form of E=mc2. So you can gauge a spin or frequency in terms of energy and mass. But I'd rather stick my head in the sand then set them out.

You know; nuclear reactions of fission and fusion probably fit in here. But who's in any position to slot them in? These reactions show a mass-energy transformation process involving motional directions. Theoretically, that could mean adjusting spherical spin to produce spiral energy. Some of which transforms into the waves of radiation. I mean your initial nuclear arrangement could be what's 'charging' the nonlinear formation of mass and motion, all of which can produce an abundance of directional motion. Through the initial arrangement you can stress the formation balance and manipulate a response. It's an automatic symmetrical adjustment which means a systematic process. So if we're going to mess around with superluminal power we're possibly looking at the system per se.

Releasing a build up allows the formation to function correctly. A certain level of nonlinear intensity, for instance, could identify with an overwhelming repulsive charge. And that could be a state of assymetry within the system. It could also add up to a proton and neutron imbalance. The thing is; the instability that particle physics determines, between protons and neutrons, could also amount to the deformation of a force. We could also be looking at the imbalance of directional motion. How precisely does it manifest? How does it release? These issues could involve formations of nonlinearity, as well as their quantum effects.

The diagram below describes a build up of the swirling 'mass of motion'. It's creating an imbalance within a nonlinear spinning spherical formation. That's the light blue ball. The light blue spirals are the consequent radioactive emission.

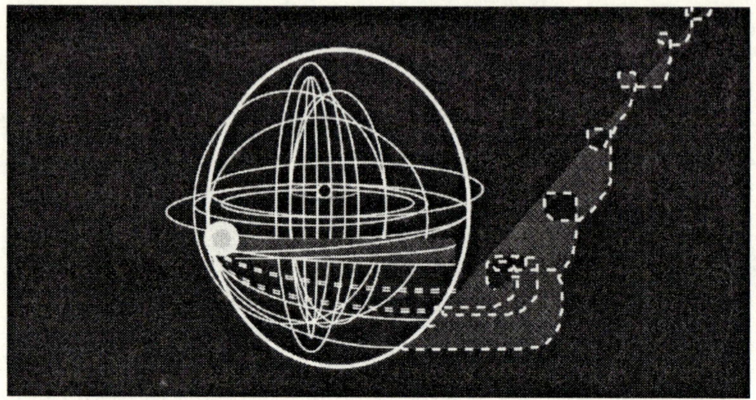

And you've still got your oscillatory charge distribution. What it means is possibly beyond me. But the name connotes the dynamics of a nonlinear situation, which is why I'm going with it. Hopefully it's the last bit of half grasped info though. Or we could end up with more loose ends than Soho. Anyway, OCD is supposedly your 'wave picture' of matter. And it's accessible through the waves of electromagnetism. How would that relate to force formations? Would it show what causes the imbalance and consequent radioactive process? Would it reveal how nonlinear systems unravel? A greater oscillation could mean a spiral trajectory carrying more spin potential. Eventually, depending on its strength, this would translate through wave formations. And there's your differentiated waves of radiation.

Initially, the charge oscillation should produce a spiral trajectory coming from the spherical spinning formation. Spirals describe the first stage of stabilising the intense nonlinearity of charge emission. Spirals space this out via elongated direction. The next stage is the wave. So the directional importance of motion—that elongation affords, comes out in the wave as well. What's the point then? I'm getting there. The spiral could figure in the wave formation in so far as the initial strength of OC emission allows. I don't want to lose the plot with this, but I think OCD could be spiralised emission of charge direction. It's strength could show how much spiral figures in the broader line of direction.

So why are some waves of radiation more susceptible to nonlinear activity than others? Some can more readily traverse nonlinear dimensions. It's your spiral imprint. Here's the point again, Harry. The nonlinear angular momentum, that transfers from spherical spin into the wave formation, would in some sense be inherent spiral formation. We can gauge just how much inherent spiral input there is by the wave's ability to penetrate matter. And the OCD factors in there. It's a sign of sub-atomic force formation activity.

You see; the spatial directions of the wave would also reduce the intense nonlinear momentum of charge. With wave you have a formation that differs in function from spiral. Again, it differs in ways beside velocity. I think it's these other wave elements that change according to spiral momentum. Look, the wave's ability to penetrate matter varies despite having the same velocity. So what, you say. That's jolly obvious, according to frequency and wavelength. The idea here though, is to accord a wavelength and frequency to spiral presence.

From the above, the extent to which the wave interacts with mass concerns the spin potential from which it derives. Some waves recoil into the spherical formation easier than others. It's tapping into nonlinear momentum. Those with greater frequency, for instance, possess more of it. That means greater spiral involvement. There's your shorter wavelengths that more readily spiral. A propensity then for nonlinearity gets the wave deeper into mass. And your radiation with longer wavelengths probably doesn't have that. Call it innate resonance. The longer wavelength, not being able to reverberate as well, overwhelms it. In short; the wave with inherent spiral is more active.

The light blue in this diagram is nonlinear angular momentum. The dark blue are the spirals and waves. We're supposedly seeing increasingly more nonlinear light blue in the tighter, smaller waves—come—spirals—come sphere of spin.

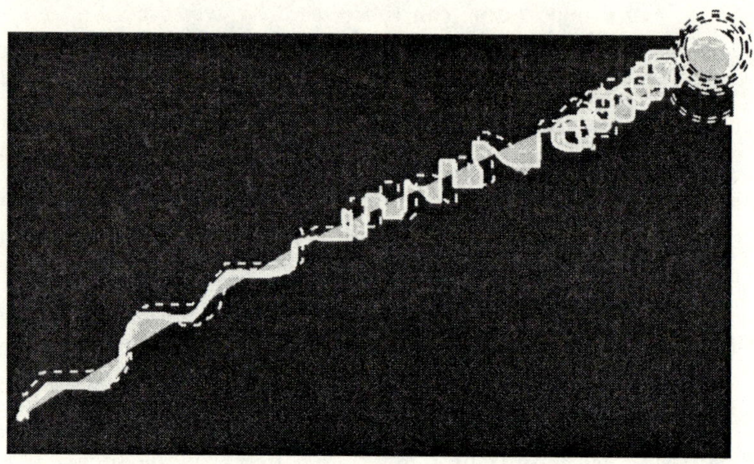

And anything we detect as radiation wouldn't show up spiral-wise with equipment subject to speed c. Here's that diagram again:

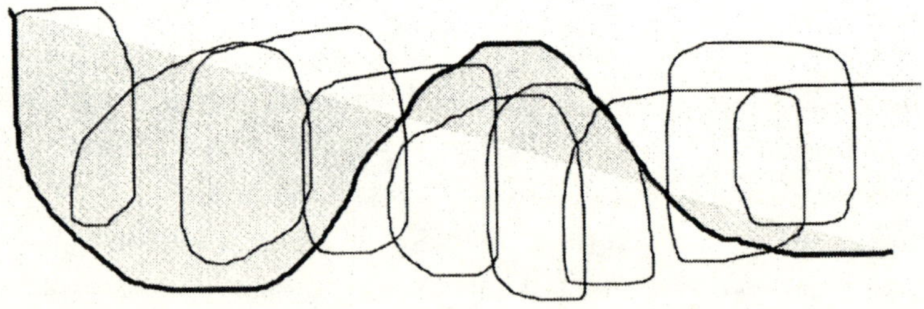

Using both red and black formations to get from left to right in the same time, means the black needs to travel faster.

Finally, our laws show that radiation can reverse time. Is this baloney, or what? Why am I sitting here with a dodgy ticker then?

Let's back track to 1980 with it. Look, I agree with Einstein's 'time dilation'. There should be an effect on time through space and motion. I just can't accept that it reverses. Clearly, the rate of decay in unstable nuclei varies from fractions of seconds to billions of years. Well, I think that depends on the balance of forces within a nucleus. The force formations and how they interact should influence the rate of decay and forms of radiation. And the force formation itself has a particular disposition towards time within space and motion. The orbital formation, for example, wouldn't have the same space, time and motion ratio as the spiral. So the internal mechanics of force formations should have some effect on the rate of decay. What a beauty. Does this mean someday we'll get cellular regeneration via spirals? Yes, well . . . the mind's beginning to boggle off the point. Perhaps our radioactive time reversal involves the same logic as what's in paragraph 1 of this section. Ergo; radioactivity seems to show how wave activity revisits its contributory force formations. Too bad I yet haven't shown it. Then again I might've. Flash back to the stuff on your sub-atomic propensity for multiple interactions. Remember, no one force predominates there.

Anyway, a forward progression of time could be one reason why there is radioactivity at all. It's addressing the consequences of the progression. Indeed, the forward progression of time should instigate the force formations. Look at the basics again. Theoretically, the primary ingredients of the universe are spacetime *and* motion. Given motion, everything is subject to change in spacetime. So you can have force formations. There's your symmetrical differentiation of spacetime and motion. These are our invariant structures. And they should be necessary to co-ordinate change. Without their interactions, imbalances of nonlinear intensity can arise. And things could randomly decay into chaos. On the other hand, motion affords a forward progression in time. It provides the scope to structure change, with a degree of order. There should be no chaos. I mean there's no necessity for chaos. We can always address any asymmetry if the whole jolly cosmos is a manifestation of symmetry. Radioactivity then, could be one of the stop gaps of this set-up. Let's call it a by-product of a stop-gap. I'm suggesting you can always have a manifestation of nonlinearity *within* a dynamic system. And that's what instigates radioactive adjustment. Simple.

Even the laws of physics show an order here. With radioactive decay you're getting invariant locations with your nuclei. That's if I read it correctly. Your emission then, could be safeguarding the force's formation. That is, certain interactions might necessary be flowing from this structural invariance. So while these processes may seem to unravel matter, they needn't be reversing time simultaneously. The force formations hold that in check with their different ratios of space, time and motion. Enter one with more space and time to address a motional intensity. Or revisit an interlude of spatial freedom still remnant within the system, to do the same thing. The consequent radioactivity, through one such forward progression in time, is preserving the invariant nature of forces.

Time's a fascinating thing. Sci-fi time travel seems to be the way to go. Roll on worm holes. I just can't get my mind around them. And I guess I'm going against the grain here with the rationale. Even so, if you go into it, theory's presenting plenty of time-space freedom. The significance of the time scale and manoeuvring force formations should be profound, according to it. Especially that is, if 2 of them exceed speed c through intense nonlinearity. These ones don't just have minimal space to maximum motion. They've got minimal time as well. I'm looking at time dilation as a minimal amount, by the way. Its density should dilute, or dilate, according to increased motion. Anyway, experiments show the effects of relativity, such as time dilation, can increase the frequency of the radiation as much as a trillion megahertz. I think here we have motion and time reactions in space through changing the shape, or pathway, of the motion at high speeds. It's getting more nonlinear. And when that happens particle physics steps in.

At intense velocities, and with radioactive processes, our laws of physics and experiments show that everything is a matter of particle differentiation. The more the particle motion curves, the more radiation it emits. Yet spirals and other force formations don't appear to figure in calculations. At least not to the extent that particle differentiation does. Maybe it's become the new wave challenge amongst physicists. Eminent recognition could just depend on naming another sub-atomic skerrick. (You'll know I'm of like mind when I start referring to spirals as lellies.) Take beta decay. If nuclei are too rich in neutrons to be stable or long-lived, they immediately begin to seek more stable regions via beta decay. This means they don't just unravel into spirals. Rather, they 'spontaneously' decompose. They go into different particles, nuclear or otherwise. Zappo. I'm not saying they don't mind you. What I'm saying is these particles are the quantum effects of a nonlinear system. There's interaction going on here, in varieties of spacetime and motion, that transcend speed c. We don't get them. We just get the after effects without the in between stuff. What a waste of fun.

Particles, theoretically, should decay and stabilise according to their force formation involvement. And that's according to four combinations of spacetime and motion, theoretically. So the time it takes for one particle to decay in contrast to another is subject to all that. Your neutron-rich situation I'm calling strong force spherical imbalance. The nonlinear spin could be manifesting neutron effects. But that's a big hypothetical at this stage. I'm pre-empting the neutrons coming up next.

These particles, as mediators of various forces or their quantum effects, only act with particular ones, it seems. Perhaps they're first off the line in terms of mass and energy transformations. As such, they could remain subject to the symmetry of the force formation. Our nuclear ones, particularly, would still link into the superluminal situation of the force. Perhaps then, those enigmatic definitions particle physics accord, such as 'strange' and 'colour', might relate to it. Remember; as particles, radiation becomes bound to the speed c limit. Whereas with force formations, there's a greater range of velocity and movement. The speed of light is just a point of symmetry. So maybe 'colour' is part of a spiral factor, or a direction of motion that doesn't calculate below speed c. The said particle could be exhibiting a nonlinear capacity that transcends speed c according to its guiding force.

I could always draw up a hypo framework linking theory with particle physics. But it's confusing enough making head or tail of the generalities. Even so, zions and pions could relate to spin and spiral. And maybe their colours and charm have E=mc2 consideration. Unfortunately, my rudimentary grasp of them pretty much equals my grasp of maths. So I'm out of here and into neutrons. Besides, some enthusiast is mowing the cement outside. What a frigging racket. There's no frigging grass around here worth plucking, let alone mowing. I was about to suggest a framework of these factors for drawing demystifying correlations. The characteristics of force formations might explain some of those definitions and quantifications of particle physics—and vice versa. Or more correctly, I was going to elaborate on it. But in this racket, bugger that for a joke.

So we're stuck with more questions. What's the neutron got to do with this? And how do the zoo particles link in with speed c squared? Furthermore, if fields modify the radiating trajectories from the atomic nucleus, could we manoeuvre them also? That is, according to force formations. Could we produce beyond speed c emissions this way? Where's our nuts and bolts of radioactive information for scientists to rev up? I think the neutron might hold some of these answers. And ditto the particle nature of alpha and beta. Alpha and beta radiation seem to reflect the contrasting formations of the negative and positive charge. So in the final sections of this chapter I'm going into neutrons and then summing up the whole kit and caboodle with charge formations.

. *neutrons*

Theoretically, the neutron is a facet of spherical spin. It's a mechanism of superluminal activity. I think it controls the conservation of motion, energy and charge within a superluminal level. So it could show up as hard core rocket fodder. And I've got the legal limits on my side with this one. Our laws describe the forces between nucleons as too mysterious to predict precise behaviour. Experiments show the binding energy of the strong nuclear force, or the spherical spinning formation, isn't cumulative. So our neutron could be doing gymnastics with asymmetrical build up. Apparently, it operates like an elastic band. If it snaps, two quarks form on each side of the split. You see; more quantum effects of notoriety. But I don't see the neutron as one of them. The neutron could be a systematic necessity, quarks or no quarks. Quark-wise it would link 'quarks' of spherical spinning formations to spiral transitions for nuclear bonding.

By providing the necessary symmetry, I think the neutron allows spherical spin to function effectively. That is, as nuclear bonding energy. Lets suppose; the spherical spinning formation comprises of protons *and* neutrons. (Although the proton still comes over as the essence of nonlinearity to me.) Anyway, perhaps when the neutron unites with the proton, it assists spherical spin to work efficiently. It could, for instance, transform and co-ordinate the motional directions of the proton's positive charge. Perhaps it's neutralising any open-ended nonlinearity into a stable framework. This could be how our spirals translate out of spherical spinning formation. This could be a treasure chest of weak force activity. Look at the angle of merger. We've got onion layers of bonding energy here. Without them, asymmetrical motion could happen in minuscule time and space. That could manifest the quarks of particle physics. Snap. Our neutron spiral layers of motion, however, could provide the bonding energy that results in nuclear stability.

They reckon the stiffest material known to physics consists of neutrons. So you might take it as the very essence of spin. But I think they've got insufficient depth for this role. Their structure will unravel and transform easily in contrast to the proton, which I prefer to think of as the axis of spherical spin. That suggests the neutron forms the sphere of spin, rather than its hub. By enclosing the proton's nonlinear intensity of positive charge, neutrons could dilute the electric force between protons. In this theoretical way they could express neutrality, denote mass in the atom and outweigh the proton.

The diagram below shows the blue proton uniting with the yellow neutron to form the brown alpha particle.

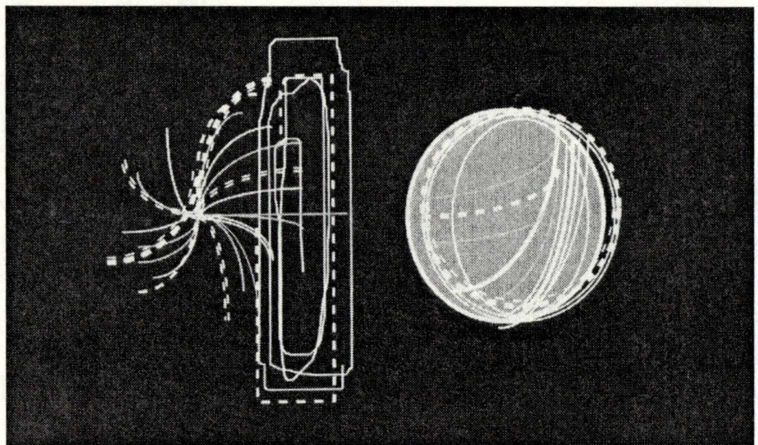

Now neutron stuff comprises of both positive and negative charge material. So you'd think it would just be a matter of re-arranging the neutron lines of force back into the relevant charge direction. And that may be the case. But the neutron mightn't have an inwards or an outwards direction. Its placement of motion in space and time could form a swirling motion at 'circular right angles' to those of both negative and positive charge directions. Such lines of force are possibly more flexible than those of the proton. So they could change direction if the situation arose. How then do they represent the stiffest material known to man? It's the bonding I think. Nuclear bond. Producing it probably ennobles them to this claim.

Besides, dislodging the spherical surface amounts to its isolation and then its decay. Physics describes this with emission and absorption of quanta. But that's stepping into the quantum effects domain. And tjhat means our neutron is no longer just being symmetrical. Particle-wise we're seeing tangible transformative effects of mass and energy. Who knows, the particle—force transition could be happening. I wonder though, that in this case it's primarily the quantised effects of the force formation. For example, the neutron will convert into a proton by emitting a particle that converts to an electron and an anti-neutrino. (The lawn mower's stopped.) Inside the neutron these would probably generate the spin. That's two down and one up. (The proton goes the other way.) That however, is according to particle bombardments. They look like manipulated transformations of the nuclear set-up. We're possibly looking at dissected symmetry with all of that. On the other hand, could a neutron unfold motion through a continuity of formation transitions?

What I'm saying, is that the world in my mind isn't just different particles. Not if stuff goes beyond speed c. I'm seeing it through a matrix of four force formations. And in the neutron part of it, we're discerning mass and energy through motion directions in nuclear space and·time. Highlighting those intricacies, here in the nuclear heart of the system, is the intense nonlinearity of proton and neutron. Perhaps this is more so than elsewhere. Again, I'm talking *directions* of motion in space and time with this scenario.

In the diagram below, is a weak force treasure chest. The pink neutron is transforming its lines of force into the blue positive and the red negative *directions* through spiral transitions.

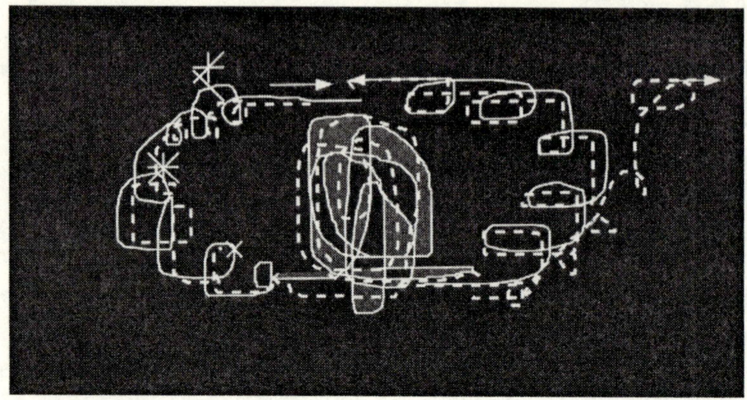

Theoretically, the neutron manages to conserve charge by aligning a spherical surface to it, thus producing an appreciable mass. That should be your mass-motion of nonlinearity though. Different that is, to what we gauge through orbital interaction. Mass-motion fusion dissects into up, down and charming segments. The thing is; the neutron shows that directions of motion don't disappear. It can transform into a mass situation. Could it prove to be an another lever of velocity in a superluminal set-up? Even in a storage capacity?

Finally, the neutron could even reflect cyclical duration. O.K. then, so I'll have it performing circus acts next. But look, a neutron's ability to conserve charge, through massive spherical formations, is perhaps a sign of the expansion or contraction phases of a galactic cycle. Without the spherical formation, protons may not bind into nuclei. They're electrically repulsion. So I've got neutrons capping the sphere on the spin. If they don't do this, protons probably decay. All up, you could be looking at a contraction phase. I'm not saying it causes the phase stage though. And I'm only speculating on the already outlined speculations. It's the neutron ability to conserve mass and charge through transforming lines of force that's the issue. This provides them with a key role in the eternal cyclical transformations within infinity. It could also be a key faculty for getting us around the galaxies.

The idea now is to manoeuvre these set-ups for superluminal velocities. And that also means controlling mass increases. Theoretically, neutrons neutralise motion by dispersing its directions. An affiliation with the weak force could control and co-ordinate a mass-motion fusion. Maybe spirals are the mechanism neutrons use to interchange their nature and identity with protons. Could that account for the binding nuclear force? We could be unravelling spin here, via a particular angle for the desired result. This sort of flexibility would countenance a diffusion of mass-motion. You see; the leverage of the neutron with the proton could influence spin velocity and spiral formation. Again, imagine removing the casement off the hub of spherical driving force. Only in this case the enclosure is a swirling motion of spherical shape and the hub is the proton motion branching outwards from a focal point. That makes the proton the spokes of spin, and the neutron a wheel of sphere. A wheel of sphere? That's a disc. Where's the thesaurus?

Spoke or spherical shell, the clue to neutron wonders might be in the formation. So the formation that charges take, especially the positive charge, could be equally important. After all, that's the formation the neutron adheres with to function most effectively. And that's that for now. Until Sooty goes for a walk the neutron can remain a gooey egg shell for all I care. The important thing here is the leash controller is on the computer while he's hanging out to climb a tree. He knows. He knows what the computer is. I showed him.

. charge formation and more facets of spherical spin

He knows what a water pistol is now as well. Climb the blinds at 3 in the morning and you get squirted, cheeky naughty. One month to go and you can play all night with the wallabies instead. So what's with alpha radiation then? Here's another possible facet of spherical spin to show up the neutron's role. Here's another clue for setting up a nuclear situation for systematic superluminal unravel. The neutron seems to provide the mechanism for alpha transportation. Which isn't to say the neutron just adds spiral impetus. A neutron framework around an alpha positive charge could form a nonlinear 'mass' of spacetime and motion. The spin gets more spherical. It's supposedly encapsulating some proton loose ends and then off it goes. Apparently, the charge and mass ratio gives the alpha particle the capacity to move at high speeds, but only over a short distance. Then it peters out. And why is that? Do those thick tracks indicate the hindrance? Maybe the casing is too thick, giving it a propensity towards mass. Maybe we're at the most agile end of the wave format where frequency teeters on the edge of spin.

Beta decay, on the other hand, doesn't release like the alpha process. Here's the need for space. I see this as too much spherical formation to start with, or too much inward pressure. This again, can mean too much nonlinear propensity to form a massive situation. Don't let me confuse you with the alpha though. Mass forming propensity here is in the initial spherical spinning set up, rather than the subsequent release. Technically, the beta results from proton-neutron imbalance. One switches into the other and out goes the excess via negative directions. So you can't, apparently, just remove some spherical surface—or channel out excess focal spin. The whole spherical spinning set-up transforms first. You've got to get the pre-balance balance right.

The resulting beta probably travels with a far more tenuous surface than alpha—in so far as our wave/particle has a surface. When the neutron changes into the proton, the framework dissipates. It puffs out into an electron cloud et al. On the other hand; when the proton changes into the neutron, it could be trying to create some surface around itself to spin out. It captures an electron cloud et al. Either way, without much spherical surface, beta could be unhampered by the condensation of alpha charge directions. Its spin is freer. It's got the space now for swirling spiral trajectories. Indeed, the negative charge doesn't seem to have mass at all. In this case, when the neutron changes into a proton, or vice versa, the excess kinetic energy that could become mass goes out as the neutrino or antineutrino. So beta travels lighter, goes farther and penetrates deeper than alpha.

But neutral gamma rays penetrate the deepest. As the excess energy from a nucleus *after* alpha or beta decay they're the result of multiple transformations of motion directions. I think that aligns them with 'non-mass' formations of anti-matter. So these formations and compositions could describe what has the propensity to go fastest. At first glance, gamma rays have high frequencies, very high energy, and short wavelengths. And theoretically, when the wavelength shrinks, the energy increases into spirals and then spin, which has a greater capacity to reach high speeds. On deeper consideration, the gamma could be akin to a high tech re-arrangement of nonlinear directions, possessing a motional memory bank of past experiences.

Often a radioactive nucleus doesn't reach a stable state through a single decay process and a series occurs. I reckon the charge twists around in the process. Anti-matter, and its changed charge, seems to result from the third sequence of such a process. Well theoretically, each transformation of nonlinearity involves directions of motion untangling and re-arranging. Nuclear stuff could end up turning inside out into diluted forms of the original version. The extent of which shows no appreciable mass content. The negative charge formation has positive charge directions, and vice versa. Through all these permutations come sub-atomic phenomena with an agility and speed that exceeds their condensed counterparts. And neutrons seem to play a key role in all of these transformation processes. As a package deal, they could hold clues to the structural formation of positive and negative charges.

Thus far I'm suggesting that a charge formation is a basic element of the force formation, and subsequent to its performance. It's the linear dimension of a highly nonlinear complex. The formation of charges involves the differentiation of one force into four—or at least the preservation of this distinction. On that basis, they would ally with spirals, orbits, waves and spinning spheres. Once realised, the charge structure seems to have a causal nature. Indeed, it theoretically here aligns with technicalities. The transformation of energy into mass with positive and negative charges defines the position and velocity of sub-atomic phenomena, apparently. So their particular role might involve the mechanics of stabilising mass and motion through defining force formations. What level of intensity triggers it off? And how do we program a charge safety mechanism into a process involving superluminal speeds?

Symmetrical invariance. Your force is functioning for and through symmetrical balance. Waves, for instance, would need both the up and down formation to flow. Orbits require two identical semi circles to rotate. It's the spherical spin and spirals that are really cluey though. But they all seem to require a balanced formation, or symmetry, to realise their force. In short; the formation the force takes should influence its performance. This really is the crux of the whole story. And there's your ostensible origin of charges.

The neutron's in on this, as well. It can manoeuvre lines of force to influence spin velocity and spiral formation. In so doing it can produce charge formations. We can see that a spiral *trajectory* can only emerge out of *spherical* spin. I'm calling it a nuclear symmetrical mechanism. The neutron helps to form a sphere through bonding lines of force. And that provides symmetry. It's spherically balancing the spin.

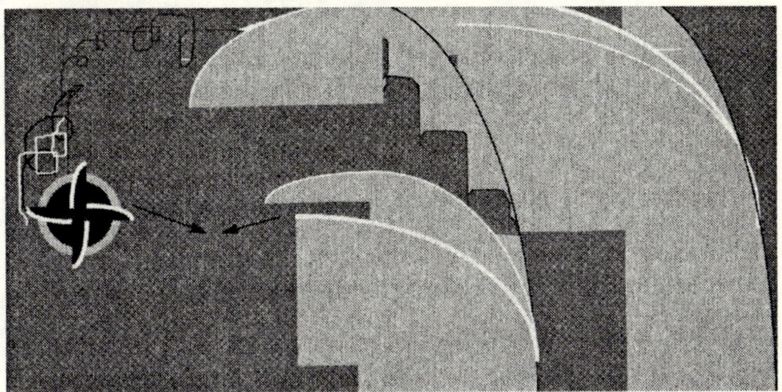

Here are some hypothetical links between charge and forces. The charge is a direction of motion within the formation. The yellow lines are the charge lines of direction. The positive are coming out of the black proton/neutron and the negative coming from the green orbital clouds. Other force formations here are: the strong nuclear force of the black sphere and the weak nuclear force of the spirals. The green sphere around the strong force formation is our neutron facet of spin and the yellow bit the proton facet.

All the ins and outs of heat, compression and thermodynamics, etc., aren't in any of that. Well, I know they're in it. But I've left them out through ignorance and for the sake of theoretical simplicity. I never said I was any rocket scientist anyway. The fundamentals of theoretical force formations and how they interact within spacetime and motion is as far as I'm taking you. Which isn't to say our laws of physics and our facts of reality can go whistle. We're just rising above their barriers in the simplest theoretical way.

Thus far then, we've got neutrons, radioactivity, and charges all organising the fastest force formation of spherical spin. The proton is beyond me. To cut a medium length story short however, it figures in the theoretical conclusion on superluminal travel. But if you want to know right now, how the spin essence of proton kicks off, forget it. I don't know. All I know is that it has to be symmetrical and the spin seems to be worth bugger all without the spherical formation. Later on, I go into cyclical end-point scenarios of macro force differentiation. That might be relevant. But it's going off this present track. Here, I'm focussing on the mechanisms of force formations, in particular the micro system of nuclear ones. To wit; the more spherical is the density, the less is the agility and the more propensity towards mass increase. Slough off some surface and go farther. But elongate your spiral into a wave and you're back to light speed and below.

Now I don't know what you made of that last diagram, but the thing is, charge formations could be indiscernible via present methods—if indeed they are *formations*. Can a linear direction, or a yellow line be a formation? Rather, their role of importance seems to involve their linear contribution to a nonlinear formation. In contrast, that is, to the particles of quantum physics. I'm suggesting the manifestations of charge lines theoretically become mechanisms of a force formation. Together with a field of some space and time, a charge—or lines of force, could then amount to a formation. And that's a nonlinear set-up through which we can calculate a rotational value.

A direction of motion, say of the positive charge, could create the field that forms the force. It's animating a region in spacetime and making everything within it definitive and active. Take for instance, the electromagnetic field. That connects the negative charged electrons to the positive charged nucleus of an atom to realise an up and down movement. There's your wave formation of the electromagnetic force. And there too, is perhaps a gross over-simplification of it, with the theoretical point presumptuously discernible. Technicalities would deny its edification, you think? But who feels like reading them? Do I feel like figuring them out? How much stuff do we have to fine tooth before we're comprehensibly in superluminal land?

Look, there should be 'laws' of symmetry. Symmetry could use charges to realise a force, according to a balanced formation. Despite all those undisclosed technicalities, symmetrical balance could be why the electromagnetic force has a wave formation. And there's plenty of ramifications with that. There's momentum, for one. To balance the positive lines of charge the electron might act to produce the wave formation. As such, the electron could carry momentum according to the wave format. That further suggests any translation of electron momentum into energy and mass—stuff that could signify as a particle position, would be within the wave format. The field accompanying the charges would only assume additional momentum according to the symmetrical need of the formation. Symmetry speaks. Now following that track, you could arrive at wave/particle issues. But I'm not going there.

Symmetry I think, rounds off the edges. We're nonlinear remember. By rendering the field everywhere radial, symmetry could ensure an even distribution of charge. Our four force formations are all round and nonlinear to varying degrees. Indeed, without the charge, symmetrical balance might not be possible. However, I think it's the other way round. Symmetry seems to provide not only balance, but the nonlinear dimension for a *line* of motion and force. After all, it's supposedly a nonlinear stage of intensity, or a level of imbalance, that sets the charge system off, thus restoring balance and force differentiation.

You know; quantum physics seems to have a different concept of formations and what they involve. Quantum physics is doubtlessly not alone in this. So the idea now is to correlate the rotational value, that particle physics place on the particle quanta of a force, to that of the force formation. Oh fun. Well, that's the idea. Reality is Sooty's slumbering acceptance of happy hour coming up.

And now that's come and gone I'm facing the unitary symmetry idea with a hang-over. Sounds like sanitation maybe, but I'm taking it as a facet of quantum theory. I think it means most particles consist of three entities. These I understand to be relative, amounting to a symmetrical formation. However, the spin factor of particle physics could also be divisible not just by particle quanta, but through the linearity of charge directions. Plus there's the formations of spirals, waves and orbits. I'm representing particles as symmetrical mechanisms of force formations. On that basis, they could involve a line of force and a field. The particle—even the charge value of the particle, could become a ratio of the force with a rotational value as per the force's formation. Your lepton then, might reflect the angular momentum of a spiral trajectory, rather than the complete spin factor of a particle. It mightn't be 'spin up' or 'spin down' but x amounts of spiral rotation. And being spiral, it would theoretically involve velocities between c and c squared.

. *summary*

Charges

Theoretically
-   lines of force
-   define force formation and maintain force differentiation
-   automatic mechanism to maintain symmetrical invariance

Technically
-   activate electromagnetism and radioactivity
-   conserve and stabilise mass

Theoretically, charges stem from a force going way beyond light speed. It possibly acts as a mechanism to maintain symmetrical balance and force differentiation. And it might do this by reducing some strong force velocity to light speed and below. It could even apply automatically to prevent infinite mass increase at speed c plus. Symmetry seems to hold the clues.

For example, an imbalance of nonlinear angular momentum within strong force spin could render the formation asymmetrical. The imbalance could activate lines of force, or charges. These could then generate fields of space through which spin excess releases. Such animation of space can also manifest the barriers of electron orbitals. A possible lever of velocity? And they're furthermore defining force formations—the most relevant one being electromagnetic wave formations. Waves, it seems, make up one of the symmetrical diversions for slowing down nonlinear momentum to speed c. An example too, of how the charge system conserves and stabilises mass whilst maintaining force differentiation?

Radioactivity:

Theoretically

- seems to stabilise the charge system
- untangles nonlinear angular momentum
- shows the active and inherent side of force formations
- reapplies force formation
- can lead to anti-matter agility

Radioactivity could also function with the charge system. It too seems to unravel nonlinear angular momentum as per asymmetrical imbalance. But it wouldn't come from the linear dimension as per charges. It possibly releases neutron *manifestation* of charge lines of force. That is, by de-activating spherical binding of spacetime and motion. These are two steps down the nonlinear ladder. And the more a line of charge re-arranges, the greater is the agility of the resulting formation. Possibly leads to an anti-matter level of agility? The agility and shape of the wave apparently reflects force formation involvement. A short wave, for instance, with high frequency signifies more active spiral involvement.

Neutron:

Theoretically

- transforms lines of force to conserve mass and charge
- forms the surface of the sphere
- akin to spherical bonding and the strong nuclear force
- a mechanism of weak force spirals

Neutrons might *transform* lines of force. But mass attributed to it should still be a nonlinear mass-motion fusion. I take it as the surface of the sphere around a positive charge. It thus bonds the strong nuclear force. It probably works through spirals. A mechanism then, of the weak force rather than a primary agent of the strong nuclear force?

With neutrons I'm putting symmetry on a legal pedestal. And these factors should be linking the linear rationale within the nonlinear. Which leads to the next questions; are we off the ground superluminally with this yet? Can gravitational mass remain stable through superluminal velocities? Can we manoeuvre force formations to have a gravitational orbital in proximity with a nuclear spherical spin?

Could spiral-neutron activity generate and maintain the spherical spin with positive charge? Could we apply an automatic charge system for transforming excess nonlinear angular momentum and prevent mass increase? And how do we generate a proton hub of charge around what might resemble an electron orbital? Symmetry is a key issue for all of that. And that means understanding the linear dimensions within the nonlinear system.

There seems to be a relationship between charges, neutrons and radioactivity. It's all part of a *system*. Specifically that is, the micro force formation one of atoms. The outlined connections could describe the balance between linear and nonlinear dimensions. I'm hoping they show how a nonlinear imbalance starts an automatic linear response. Working within a nonlinear system though, the linear would again differentiate through into other nonlinear formations. And it's all under the guidance of symmetry.

Neutron control of positive charge and the spherical spin, as well as the negative clouds, bridges these dimensions. They involve lines of force as well as spin. Neutron spirals seem to transverse both the linear and the nonlinear.

Finally, radioactivity shows a further re-arrangement of charge lines through symmetrical diversions. I reckon it can produce anti-directions of motion with increased agility. Can we use this to direct charge lines accordingly? There's no answers to any of these questions straight off. Well, not by me, anyway. They should unfold however, as we proceed. Ideally, everything here is supposed to culminate into the logistics of superluminal wizardry. Failing that, and more likely, I could stumble upon some relevant pointers for the ride.

In the next chapters I'm going deeper into conceptual development. We're out of the atom, at last. Far out. I'm going into alpha and omega again, although it's not all philosophy. Symmetry and the nonlinear with the linear remain constant guidelines.

what other way have we
but to mark when, and where, eclipses be?
to take a latitude
sun, or stars are then viewed
at their brightest
but to conclude of longitudes
how great love is
presence best trial makes
but absence tries how long this love will be

Who said this? There's the brilliance of another jolly lost referance.

. endnotes

nb 1 how I understand synergy: the working together of 2 or more things to produce an effect greater than the sum of their individual effects—ref. the Collins Dictionary

ch.6

a cosmic code?

. preamble

Now we're into deep spacetime and motion. The nonlinear theory is going into the greater context. So we're doing the whole cosmos forthwith. The idea is to see how nonlinear systems structure reality according to eternal infinity, which entails taking force formations to the bigger picture. If indeed they do, that is. That means there's some philosophy coming up. It also means teasing out the point of motion within the spacetime concept. How else are you going to get recurrent cycles?

Then we're full on symmetry. Well, full on according to its structural application within an eternal and infinite cosmos. We might get a symmetrical order out of chaos. Contemporary cosmology tells us little about symmetry, at least it hasn't told me much. There's nothing around on four forces arising symmetrically out of one. And I haven't come across any symmetrical explanation to the cosmos, cycles or otherwise.

Four forces arising out of one should also concern mass transforming into motion. We're unravelling its nonlinear fusion, professedly. The nuclear spherical spin, or our fourth force formation, is the main subject here. A tangle of mass and motion shrouding neutrons, protons, and charge lines of force, maybe—but there in lies our superluminal hopes. On that basis, there's a section on mass-motion fusion.

Finally, Einstein's genius arrives. And well you may think it's been sadly lacking thus far. I'm going to directly address the equation, E=mc2. Thinkable finally blinks at reality. No maths mind you. We're pushing it as it is. I'm only good for suppositions, and there's buckets of those coming up. Even so, the plan is to tie theory as close to reality as possible through this equation. So we could end up with more stuff on particle physics and the force formations . . . and a twisted version of Einstein's genius.

. infinite cycles within eternity

Cosmology is probably still open for plenty of theoretical supposition. Who knows what value is the Hubble constant. Why can't we measure the universe with an agreed precision? Perhaps the answers concern science itself, as much as the size of the subject. It all depends on how much you want to understand, I think. An eternal past and future should, by definition, be incomprehensible. Except that is, through theology, deities or perhaps even through Plotinus' unknowable knowledge. So there's conceivably much nobody understands. Our known universe is younger than its oldest stars . . . someone said somewhere. Try and figure that one out. And nothing amounts to a big bang?

And what's the point of extrapolating with human experience? I tried it once in high school history. I figured that the earth could be analogous of convict Australia way back when. And our neighbours in this solar system could be strategic reminders of planetary warfare. Armageddon of days gone by. Sure, that might reflect the thinking of a juvenile delinquent. But going by our history, it can look pretty likely. Far safer then to rise above empiricism and ponder the greater galaxies through divine inspiration, you think? Or just take it as it comes through Hubble's telescope. Or don't think about it. Put your head in the sand and let human realities overwhelm you. Well, I might if they weren't so dangerous.

Anyway being woebegone is a waste of time here because there's definate hope in this rationale. A cosmos subject to change and transformation could be a constant transcending our observation. What takes place, and how it happens could concern a structural mechanism beyond our aggressive grip. On the greatest level, that of infinity and eternity, the only mechanism I can think of that meets these criteria are symmetrical ones to which cycles conform.

To some extent, cycles might even confirm a theological reasoning within the scientific arena. Or maybe not. Yet beyond—and within—the scientific walls of reality you've got divine guidelines set down to regulate transformation, in terms of human evolution. Be comforted, for instance, you and they shall live again through the conservation of energy and the eternal return of cyclical transformation. And there's no tongue in cheek about that . No irreverence intended. I'm quite serious. In light of human history, you'd think divine principles would get as much attention as the scientific ones anyway. Heavens above they're based on eternal and infinite wisdom, ostensibly metaphored and handed down through the ages. Too bad for us the only thing of biblical interest these days seems the stuff about the war to end all wars. We've had it. It has become *them*. We've had them in the bible, and we're still having them out of the bible. And all you're getting is more inbred aggro, more chronicles of them and smarter weapons. And if we don't stop we'll blow those up as well. And still they're going at it tooth and nail over there. How many more do they frigging want? Neighbourly love? You must be joking.

Cycles show up all the time. Even Darwin's theory of evolution has cyclical time in cell division. You've got your rise and fall of civilisations right through human history. Economic cycles, hedonistic cycles. Indeed, both the reality and the idea of cycles is simple. Cycles show up as a measuring device applicable to everything. The challenge is to work out the structure of cycles within infinity and eternity—which brings us back to the nonlinear dimension. I mean how are you going to have linear cycle?

Mind you, a linear relation is possibly the easiest response from observation and experiment. It's the only mathematical form where every observer on every galaxy gets an identical velocity-distance relation, evidently. It adds up to the expanding universe theory. But does this really explain the whole picture? How do we know our observations aren't within a deeper cyclical process with nonlinear proportions? There's stuff out there we clearly don't see. Look at the dark matter, for starters. Our laws of physics also suggest it by their limitations. Our mathematics suggest it by their infinite mysteries. And the nonlinear variety only seems to confirm our cosmic incomprehensibility. So says the mathematical midget.

Besides, no one reasons anything beyond speed c. That's going by what I read and I don't read science fiction (hope I'm not writing it). But the thing is, for all we know, the lifetime of some atoms depends on how far back in time we can see below the speed of light, on the linear plane.

Through the linear plane, and our powers of observation, we can bind all points to a common origin at some time in the past. There's your big bang scenario again. The linear rationale seems to predict a past, all time, creation event. And one such event, for all I know, could've been the easiest way to get a moral creed across on evolution, anything else being beyond belief, too complex and too distant. So I'm wondering that today, the big bang of a linear rationale is a) of no moral value; and b) doesn't explain reality. First, your moral value of a cycle could rest to some degree, with the observation and measurement contemporary society demands. What goes around comes around. With your big bang, what goes around doesn't come from anything. The realities of a big bang put the blockers on continuity. And if you take that along the path of moral reasoning you could arrive at the loss of consequential reasoning. Why be concerned with the consquences of your actions when everything comes from nothing anyway? We only consider what we see and guide our actions from that perspective. And secondly, it's probably denying some deeper, broader cosmic dimensions.

Take galaxies. For us, there's no apparent edge to their distribution. I mean they seem uniform on the largest scale as we know it, throughout the cosmos. Now I reckon that's the big bang time scale. Which also means we could be measuring them through the linear rationale. Even so, we know they come in clusters surrounded by gigantic voids. So how do we know these inter-galactic voids don't relate to some cyclical perimeters of a nonlinear dimension? How do we know they don't concern the ins and outs of an eternal and infinite cosmos? Hell no, says Hubble. They're all expanding, as far as the eye can see. There's no ins and outs about them. I suppose he'd say that.

I'm just thinking the universe interconnects in a deep, and as yet not perceptual way, on a level where motion is within space and time. I'm suggesting the structural nature of the cosmos has a primary component of motion, and when we put that into the spacetime continuum we can draw nonlinear relationships. It takes us out of the linear plane in more ways than one. It describes forces curving and functioning through different dimensions. The nonlinear variety reflects a symmetrically invariant pattern of nature. And this seems to structure transformation. Here then is an instrument of extrapolation. You see cycles are symmetrical. Your symmetrical invariance could function through your cycles. And their symmetrical arrangement can figure throughout eternity and infinity.

Moreover, the nonlinear dimension of motion in spacetime opens the door for time to progress, rather than back track. It's allowing time to be eternal. And eternity affords endless cycles of change within infinite space. So stuff needn't go backwards and forwards. There's room to move. Why should everything regurgitate or blow up from nowhere and nothing? Phenomena can circulate spherically anywhere forward according to the cyclical patterns of the cosmos. Everything, furthermore, would be different. Nothing could be identical. How could it be with the necessity of motion and its room for change? And that's even if everything involves the machinations of previous and future things.

Based on the spacetime concept, however, the present laws of physics allow for a time reversal. Newton's mechanics, Einstein's relativity and the quantum mechanics of Heisenberg and Schrodinger still seem to function with time running in reverse. So it's no easy task to suggest otherwise. Even so, who could deny that our laws and theories are incomplete? You probably couldn't go one entire cycle with them.

And you know what; there's possibly more chance of going to heaven forever, or forever burning in hell with your linear noncyclical rationale anyway. You'd get to the end of the line and that would be it. On the other hand, where time is irreversible, the same thing never recurs exactly. It's always moving on subject to some aspect of transformation. (That means there's no two Jesus Christs, Guiseppe.) I mean you could get to hell, fire away for a while, transforming your matter with the talents you've developed until you show up in some cosmic nook and cranny as a goanna. Or you could make it to heaven, shoot the breeze and become some new age superior. The point however is; through a nonlinear dimension you get an association between the temporal and the eternal.

The tides, seasons, solstices, and astral movements all reflect a cyclic pattern of time in a forward progression. Cycles incorporate change necessarily within a recurring structural pattern. As Aristotle would say 'there is a circle in all other things that have a natural movement and come into being and pass away. . . .'

Remember, the challenge is to validate and use superluminal speeds. I think it's only through extending our laws of physics we can do this. They seem to function and explain processes only up to a certain stage. And we seem to rebound from their limitations. Cycles represent continual order. They can make sense out of chaos. But what we see through our present laws is increasing entropy. Where's your structural invariance of symmetry with entropy?

Black holes for instance, according to our laws of physics, can close off a system. That's also where our laws break down. So you don't see a complete transformation of energy through them. On the other hand, black holes could signify a major symmetrical transition. They could preclude the sequence of force formation unification and differentiation. Galactically, they could apply to a concluding phase of a cycle. Or on a different level, they could refer to the stellar system. Apparently, the lifetimes and masses of stars depend on a ratio of different forces. There's your stellar system then. If so, a stellar black hole could figure in its systematic transformation. It too could reflect the changing order of four forces. But that's for later. Now I'm trying to clarify the extrapolation possibilities of cycles. They reach well beyond the boundaries of our present laws. The symmetry and systems within them could shed a different light on many mysteries confronting us. Entropy, time reversal and black holes are just some. There's plenty of others.

. *a cosmic code?*

The farther out we look into the universe the further back in time we see. And the less detail we find. It's a meagre fragment of physical reality. So what about the greater reality? Doesn't that have meaning, if not to us then to someone smarter? I bet Plotinus would have it meaning something, albeit something incomprehensible. I know, I read him at university. "Is Plotinus' Theory of The One Coherent?" Well no, but you had to stretch it out to 1500 words. And in restrospect he might be more lucid than mud to others. Anyway I think we can work it out—the cosmos if not Plotinus. We should be able to extrapolate beyond the linear plane of entropy and chaos. Turn on the superluminal light. Get out of the solar system. Bounce off the galaxies. Hang about in the Andromeda Spiral for a while. Check out the Magellanic Clouds. Wander round the Milky Way. Then leave the neighbourhood altogether. There's probably stuff out there—places to go, phenomena to meet—we can't begin to ponder here in our bubble. There's probably even stuff going on in our bubble we're not pondering properly—Neptune for one. And I'm thinking there's a framework transcending our linear one. And it's the nonlinear cyclical framework. And it should encapsulate formulas and deductions for getting out there. Simple ones, through which we can extend our laws of physics and break out of the bubble.

Evidently we have yet to crack the code of eternity and infinity. The spacetime concept and divine belief seem to be the closest we've come. Spacetime clearly explains a big slab of it. But spacetime *and motion* might cover the lot. What I mean is we should be able to carry the spacetime concept into a bigger dimension. $E=mc2$ seems to apply to time, space *and motion*. Here we have the motion of c2 which, according to the linear frame of spacetime, stops at c. So I wonder that it can apply to plenty of stuff beyond the spacetime framework. You see; part of cracking the code could mean grasping superluminal velocities. You can go the distance then, physically and intellectually. It seems to add up every way you look at it. The fantastic distances our cosmos provides are meaningless to us until we first grasp them intellectually, and the superluminal factor appears part of that process.

Motion in the spacetime concept provides the circularity of multidimensions. Through it you find formations showing structural invariance. They're not chaotic and describe patterns within nature on all levels. Furthermore they appear to describe past and future processes. Look at the spiral arms of disc galaxies. They signify areas where young stars form. But they also relate to a wave pattern in the disc with forerunners showing regions of even greater density. These formations figure right down to the sub-atomic level. On the bigger scale though, they could amount to a cyclical framework.

Mind you; I'm taking cycles *within* eternity and infinity as the maximum systems of spacetime and motion. I'm not talking cycles *of* infinite and eternal spacetime and motion. Nor am I suggesting mini and maxi cycles. I think nature's basically simple at the bottom line. And the symmetry of cycles comes over simplest to me. If you pattern infinite space and eternal time in cycles you're also drawing in symmetrical motion. For this to work; time and motion would be as inseparable as space and time. Indeed, they all should be impossibly inseperable. But the real issue here is how motion slots in. Does it shape time in space, or space in time? Well, none of them would come first in an eternal and infinite cosmos. So it's likely just a matter of how they affect each other.

Theoretically, these are our universal constants. Spacetime and motion could be the irreducible structure of everything. The forces symmetrically express this. I mean they have no independent life of their own. What's one without the other—an overwhelming black hole of linearity? The total imbalance of chaos? Transformation gone wrong? For us, forces only seem to create the order of spacetime to a certain stage. Or at least we only explain this order to a certain stage by way of spacetime. Well, I think forces are a systematic differentiation of spacetime *and motion*. And together they can reveal a structural business of spacetime and motion, one that's based on balance.

Through the present laws of physics, however, motion seems to be an indirect consequence of space through time. Sure, the nature of spacetime might remain valid according to its general relativity below speed c. That's just your macro gravitational effect. We're judging it according to a level of motion. Why should spacetime become an inextricably bound phenomena from motion? It sure doesn't come over that way with the nuclear forces. When do we judge motion and time according to space? What happens to space and motion, according to time? Three dimensions of space and one of time don't specify which way it flows—forwards or backwards. You need motion for that. Motion should be a decisive factor. How can it be otherwise when at the speed of light everything is unknowable? But that's your linear level for you.

To account for motion's position in the spacetime concept I think we need to define the flexibility of motion and time. In particular what reversible mean. Evidently it's an issue that keeps cropping up. So it could account for a fair whack of bubble vision. Well theoretically, motion goes in any direction and at any rate. And time, according to this theory, also marches in the same direction as motion. So does space, for that matter. In theoretical fact they're all inter-connected. Reversing time then would also be reversing the space connection. One cannot exist here without the other. It's that simple.

So being subject to each other, more motion can mean less space and time, etc. And it could be on that basis dark matter figures. I don't know about worm holes, though. In cracking the cosmic code we're going to worm hole around the stars? Speak for yourself. Where's your style? Anyway, dark matter might amount to the room for recycling. More than likely it should be a bygone manifestation of spacetime and motion. Next door's available space, for instance. Even with eternal and infinite possibilities I don't see how you'd get any newly minted spacetime and motion. But the elusive presence of dark matter could account for the structural links of time, within space and motion of *cyclical scale symmetry*.

Too bad I can't name any galactic contraction phases in the cosmos as hard evidence. But what's with the different galactic shapes anyway? Do they fit into some macro system of four force formations—one that reaches cyclical proportions? I don't know. It's all just speculation. And it could all remain that way because here we are, below the speed of light, looking out on an expanding universe. Nevertheless, without even speculating an order of motion through an eternal and infinite spacetime, everything just assumes the linearity of a closed system. And we could remain here, stuck in a rut of reality.

Let me go over it again. Metaphysical and abstract it might sound, but it could hold some tin tacks of greater reality. Per se, spacetime and motion are equally relative. Taken individually, their strength is relative and depends on their combination. Indeed, you can't take them individually. But you can get different combinations of them which amount to a force. Even according to the present laws of physics, time progresses slower where motion is faster. More motion then, less time. And where time reduces it *dilates*, along with space as motion increases. But that's as far as speed c. That's as far as spacetime can stretch through motion on a linear level as per greater gravity. Then it starts to curve in on itself. That's when we're going nonlinear.

The point is; time never reaches zero. Not in your eternal world with necessary motion, it doesn't. Rather, speed c motion becomes a factor of invariance with a structural symmetrical application. The impact of motion here is only reducing time's effect. It's not completely overwhelming it. Certain amounts of each signify a symmetrical stance. They describe a formation and a force. And you get decisive transformations of mass and energy through them. Speed c might be the end of the line, but it's not the end of the story. Along with discrete amounts of space and time, it just figures decisively throughout the eternal and infinite *transforming* fabric of the cosmos.

I think motion in spacetime becomes a causal phenomenon with levels and strengths. But like spacetime, is not a total cause. Instead, at four levels one of which reflects speed c, their combination realises a formation that acts as a force. In other words through time, motion can animate space into the activity of a force. And these force formations seem to represent invariant symmetrical structures of systems within the cosmos. They effect transformations of mass and energy accordingly. And any more of this, I'll be affected as well. Come on Sooty, wake up, I need to sit in a tree.

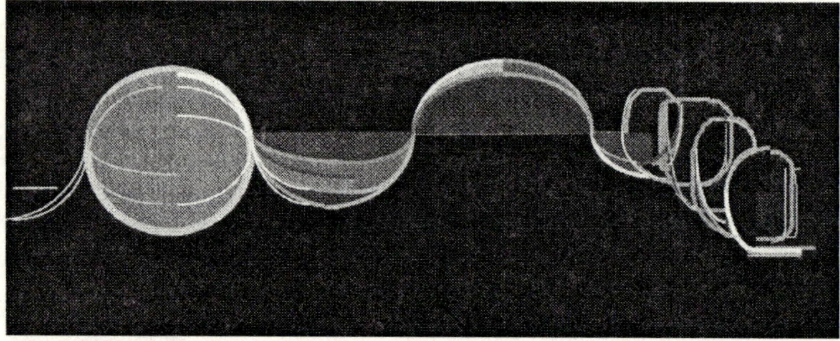

The diagram above shows the linear format of timespace and motion reaching full circle on the left. Time is the light blue circle. When it comes full circle you'd think it would be zero or reverse. Motion, however, transforms space and time into a different formation. With the wave, we're getting more motion and less space and time. And everything curves more. Time doesn't stop at the limitation of the gravitational force formation. It's still marching forward. Diminished and dilated, it's now in a different combination of space and motion. It's thus expressing another force. See the green and yellow time dots. Furthermore, the wave formation shows how time can increasingly link with space through the motion of energy as well as matter. Here energy, as well as matter, seems to be curving space. There's your nonlinear fusion happening. With more motion and greater nonlinearity the formation can enact mass and energy transformations seemingly concurrently.

Following that comes the pink spiral. That formation has even less space and time, with motion between speed c and c squared. It could also mean increased concurrent transformation possibilities. What a mouthful. The outward signs of mental over-load, no doubt. Time then to turn the computer off and check out the end of the midday movie. And it's a ripper—'Misery' with James Caan. She makes him burn the only copy of his unpublished novel in it. Missed it. It's frigging over. So the point then is this; time, dotted above, advances from light blue to green and then pink spirals. Progressing in a forward manner, it travels along with the nonlinear curvatures of space and motion. And it's all because it necessarily fuses with space and motion.

Look, despite the invariant nature symmetry imposes through these formations, spacetime and motion are malleable agents. Given the subsequent energy and matter transformations, we could bear this in mind. Somehow, through a *system* of mass and energy, ergo a system of 4 forces, we might be able to maintain time on the level of a gravitational orbital, and still allow for the manifestion of motion at speed c squared. That's not a worm hole. That's superluminal travel. And you think that sounds improbable? Imagine the jolly equations. Well, don't. Your brain might explode. Mine would.

. *Symmetrical Invariance and 1 force into 4*

If someone asked me (in my dreams) what the code of the cosmos involves, I'd say the symmetry of spacetime and motion. And that's no secret. Symmetry is everywhere. It's not as if you have to dig a mine to it. What I mean is, how the cosmos works is probably no cryptic mystery. It could be a simple open plan for anyone to work out, should they so desire. Except, that is, mathematically. Symmetries, and the invariance they reveal, reflects laws of nature all the way down the line. Moreover, they're nonlinear. Where's your chaos when there's symmetry? You've got a balanced system instead. Technically, whilst the balance remains, changes can occur without altering essential features of the system. It's either that or turtles on plates with stuffed monkeys.

Looking at things symmetrically, we could identify a structural application within phenomena. We could perhaps find the invariance that creates order out of chaos. Even on the most macro level within the infinity and eternity, you'd still have symmetry. I reckon cycles are a form of symmetry. Endless expansion isn't. Any idiot can tell that. I for one, can tell that. Where's your pattern of balance? At some stage, you'd need a contraction to get one. And there's the makings of a cycle. The whole cosmos could be subject to phases of expansion and contraction, just like my lousy figure. Its nature might continually apply within infinity and eternity. Endless fatties and skinnies. Whatever. Just don't call me fat, bone head.

Anyway, I think the bifurcation of symmetry could occur through these cycles. That means you get systems within systems. Basically, bifurcation here concerns the differentiation of spacetime and motion into four forces. Well, these forces could further bifurcate according to a symmetrical paradigm. Under your cyclical umbrella this could be nature's way of expressing universal constants with local ones. Remember, we have spiral galaxies as well spiral radiation. There's even spiral patterns found in shells and plants. And what's a tornado anyway except funnels of wind going faster than is laterally normal—according to system. Our spiral formation is the theoretical formation of the weak nuclear force. Coming from the spherical spin, it goes into the wave and then the orbital, which all amounts to a system of forces.

Symmetry could be why there are forces in the first place. With them, we have the order of symmetry within the cosmos. Here we have four configurations of spacetime and motion that describe a structural mechanism of balance. All of which applies within transformation processes. And that's on various levels throughout the cosmos too. Indeed the force, teleologically, could exist to maintain invariance. As an expression of symmetry they function for that purpose. What they manifest reflects it. But get this. Going by the above, the symmetry of them also reflects a causal mechanism. And it's relevant to infinite and eternal spacetime and motion. Symmetry fashions transformations such that there *is* no final cause. Theoretically, that is. One force transpires into four. And the four can transpire back into one according to the symmetrical framework of cycles.

These force formations should demonstrate how the laws of physics work through symmetry. They're four simple patterns showing an order within the cosmos. And sure, just putting forward circles, waves, spirals and spheres is no big revelation. It's probably only stating the obvious through presumption and mathematical ignorance. I guess it depends on your cosmic focus though. The right one should enable you to distinguish between bunkum and possibilities. Either way, these things seem to describe a configuration of spacetime and motion that is highly symmetrical. Further still, they apparently apply as the most efficient means to transform energy and matter.

Take charges. Do they describe a symmetrical application of motion within forces? I've been suggesting charges activate where there's an imbalance within the system, most noticeably the strong nuclear force. This imbalance could be your nonlinear motion build-up within a spherical spinning formation. Perhaps the charge system is a means to symmetrically distribute it. The fields of waves and spirals outside the sphere of strong force reflect it. When it's radially animating space thus, symmetry could be securing an even distribution of charge. Indeed, these fields outside the sphere could amount to a concentration of charge within a central point. And that's balancing motion within space and time through the differentiation of one force formation into others.

So where does one force come from in the first place?

Theoretically forces are just a manifestation of spacetime and motion. And they occur because of the dynamic element motion presents within space and time. I reckon the symmetry of a cyclical pattern initially differentiates them. Cycles surrounding cycles would affect each other's spacetime and motion. A contracting cycle could unify spacetime and motion into one force. Say that of a tightly knit ball of strong force spherical spin. You could get a big bang out of it, kicking off the next sequence.

I'm only about theory, you know. If it's coming across as know all abstraction, it's because I'm getting familiar with it. Anyway, our cycle is a symmetrical occurrence. It embodies symmetry. Each cycle is the product of another and subject to neighbouring ones. You see, my position philosophically has always been: that the universe interconnects in eternal transformation. And this theory of cycles deduces from that conviction. Your cycles could *effect* each other. Given an invariant nature, according to symmetry in motion, their expansion and contraction rates are also a balancing act. En masse, they balance the properties of spacetime and motion *within* each other as well as externally. So the big bang of one cycle could adjoin, and encourage an 'omega' stage of another one. In other words, a cyclical juncture of force unification could prompt next door's expansion phase changing into a contraction one. And it's all for the sake of symmetrical balance.

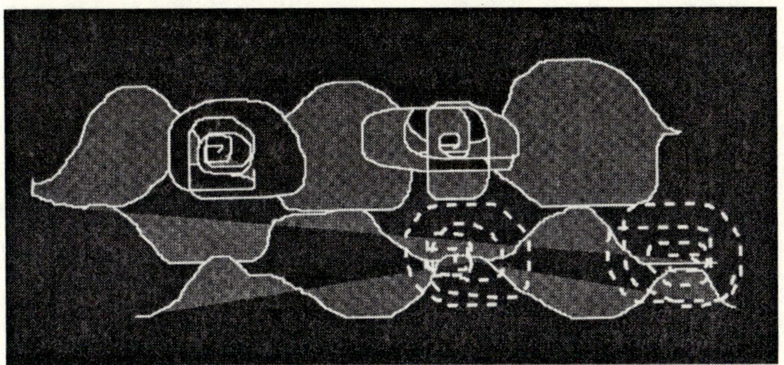

Here we have cause and effect on the biggest scale going. Your stellar and galactic systems could amount to one massive system of spacetime and motion. Call it the cyclical scale differentiation of four forces. And this is where the general relativity of a force can predominate over sub-systems of them. Nor is that the end of the story. There's no end. There could be loads of them. We're eternal and infinite remember. The condensation of some part of it can always provoke, contain, and control, the differentiation and expansion of another bit. In this way, cycles could continually manifest through their own pattern of symmetrical invariance.

Even so, we're not talking simply space here. What ostensibly contracts and expands is spacetime and motion. That's how the invariance of symmetry functions. You've always got a ratio of time *and motion* within space to define the force on any level. In the nonlinear atmosphere of cyclical dimensions, one force of spacetime and motion can explode, with a big bang, manifesting four formations and bifurcating into sub-systems of them. I guess that's throwing a spanner in the works of general relativity. Where gravity's prevailing in one neck of the woods, your nuclear forces could be overwhelming stuff in another. We're going beyond macro gravity, supposedly with this scenario. I'm into the macro dimensions of systematic force differentiation—but more of that later. For now, there's more spirals, symmetry and space/motion coming out of my ears than I can handle. Where's that Mr. Whippy? He usually shows up about now. In the meantime then, will spacetime and motion cease? In and out it may go. But I doubt that it's ever going to fizzle out, or freeze static forever whether I'm plugged into Fred here or not. Nevertheless, the challenge remains to find out. The stars keep on twinkling provocatively each night. They're up there and we're down here. And I'm full on with this for some time yet. How's it going to happen anyway—unless we get on with it?

. *mass into motion*

Force formations also describe transformations of energy and matter. That's their point. With them you've got the order of symmetry all through reality. Look, the four forces technically create order out of chaos. We're just taking that proposition beyond our observational boundaries. And is it really so abstract? This is no ideal. Cycles of symmetry isn't wishful thinking on my behalf. Sure it's full on theory, but I'm still trying to extrapolate reality according to it. Well, admittedly it's one hell of an extrapolation. This symmetrical pattern of cycles—what's 'conceivably' generating forces of spacetime and motion, isn't just out of reach of Hubble's range. We're talking eons of eternity here. But for all we know we could already be seeing some of them. The pattern of symmetry supposedly bifurcates; remember. It could thus cause systematic differentiation of these forces. And you can still find that within a Hubble range. I think Anaximander reasoned stuff along these lines. He had a 'primary substance' that I'm reckoning as spacetime and motion. It's supposedly infinite, eternal and ageless as well. Plus it transforms into familiar substances. So maybe his familiar substances could arise through these patterns of symmetry.

Two paragraphs, Sooty. That's all I've done in an hour. You want to come in. You want to go out. And you're not jumping over into the neighbour's back-yard again. Bloody wonder you didn't hang yourself last time. You want to play games—that's all you want.

Well, I'm not playing last night's one again. Look at me off the leash. Can't get me under the parked cars. Can't get me right up the tree. And there you were, ready to jump over the back fence and check out the train tracks, like those birds did. That'd be right. Splat. And then I've got a big heart ache. Thank heavens we're on the count-down out of here. The only high-ways, by-ways, traffic and sprawl I want to see or hear is what's heading south to the boat.

In the meantime we've got a wave, orbital, spiral, and spherical nature to phenomena. They all have different combinations of spacetime and motion. We go from the minimal spacetime and maximum motion of spherical spin, to the maximum spacetime and minimal motion of the gravitational orbital. I've gone through that already. The thing is; the percentage of motion with spacetime affects the degree of curvature of spacetime. That's your force formation happening. The spherical spinning formation of the strong nuclear force shows space and time at a minimal capacity and motion at a maximum. And that particular framework has so much curvature, or intense nonlinearity, the formation seems to fuse the motion of spacetime into solid spin. It seems to encapsulate the motion of speed c2, condensing it into mass for lack of space, time and linearity. In the big picture I think this one figures at cyclical junctions.

Now any mass situation attributable to one such formation, say of the proton, should involve this nonlinear pattern of total curvature of spacetime and motion. What then do I mean by 'mass situation'? Well, I don't mean scrambled eggs. Again I'm talking solid rotation here. We're at the point of maximum condensation. In this way it's perhaps analogous of what precedes the big bang, or what comprises of cyclical end-points. But that's not the point.

On many levels these formations generate transformations. It's a matter of the force's degree of nonlinearity. That's what determines your mass/energy ratio. Where the formation allows for increasing motion within decreasing spacetime, you've got more nonlinearity. More nonlinearity might mean more subsequent energy. Supposedly it means more mass-motion fusion. Remember though; they don't act individually. And I hope you're still with me on this. We've got enough plots on the go to rival Shakespear. I'm up to draft number 4 and still paring off nuances. Anyway, your mass/energy ratio is in some sense contingent. It's still a systematic thing. It's still a symmetrical derivative. You see; the force formation remains a structural component of a symmetrical system. No single force is per se symmetrically valid. It stays contingent upon a metaphysical unification with the other three. And I'm saying that, regardless of Fred's reading rating. (Fred the computer here, likes his words and sentences short and sweet.)

Looking at them individually however, you can see that one formation allows mass and energy to transform through it's unique dimensional pattern. That is, though it's own degree of nonlinearity. Your wave, for instance, is less spherical and more linear than a spiral. With this formation nonlinear motion gives way to spacetime. Indeed, the wave seems to debut the nonlinear mass-motion fusion—or your 'mass situation'. It's on the cutting edge of nonlinearity. It has neither the spin of the spiral, nor the linear sweep of the orbital. The effect is a wave force at speed c. And according to E=mc2, I think the wave concerns equal amounts of energy and mass. Theoretically, here we have energy and mass transforming in and out of each other at speed c. Other formations should produce different ratio's of mass/energy.

And what about the slowest end of the force set-up. The gravitational formation describes a lot of spacetime with hardly any motion—at least hardly any of the nonlinear variety. The effect is an orbital force below speed c *surrounding* mass. Indeed, the source of gravity becomes mass that equals its measure of inertia, or 'absence' of motion. But it's in there. In one form or another motion is in everything here. I've got motion on the brain. And how and where mass figures—well at this stage I'm not going to unravel it according to a system. Do you want to get really confused, or what? Remember; the macro one of general relativity gravity doesn't tie in with the electron one of atoms, so simply. Let me just say that the motional content in the spacetime mixture of the force of gravity is minimal on every level, and because of that you can locate a mass position. But on every level these four force formations should act together to produce mass and energy transformations according to their system. You've got stars and planets, for instance, in the stellar one. And you've got your electrons and photons in the atom. Their bifurcating systematic repercussions is another story.

The immediate, previous point is that there's never an entire absence of motion from space and time. It doesn't matter what form it takes, orbital or spiral. Yet the orbital looks linear—without any nonlinear mixture of mass and motion. The symmetrical relativity of these force formations, however, requires gravity to be nonlinear. Why then is the motion of gravity, within spacetime, practically indecipherable compared to the other three forces? Hark back to the systems logic. Are we comparing the *macro* gravitational pull of general relativity with the *atomic* wizardry of nuclear forces? At present the grand nonlinear framework is beyond the borders of gravity as we know it through the laws of physics. According to our present laws of physics we only reason it through an expansion of the universe. (And on that basis, anti-gravity may not be such a bad idea.) You see; gauging a macro orbital on the cyclical scale could render it nonlinear according to a contraction phase, and contingent upon a system that includes a 'gigantic' spherical big bang (it should spin everything down into a microcosm). On the other hand, your macro gravity alongside the micro set-up of an atom is hardly curvaceous.

You know I still want to link force formations with the transformation equation, E=mc2. It'll be a shemozzle without clarifying the mass-motion confusion first though. So I'm going back to basics for a while. Two things are theoretically a priori. First, energy/mass transformations depend on the *formation* of motion as well as spacetime. And second, motion can fuse into mass through nonlinear spacetime.

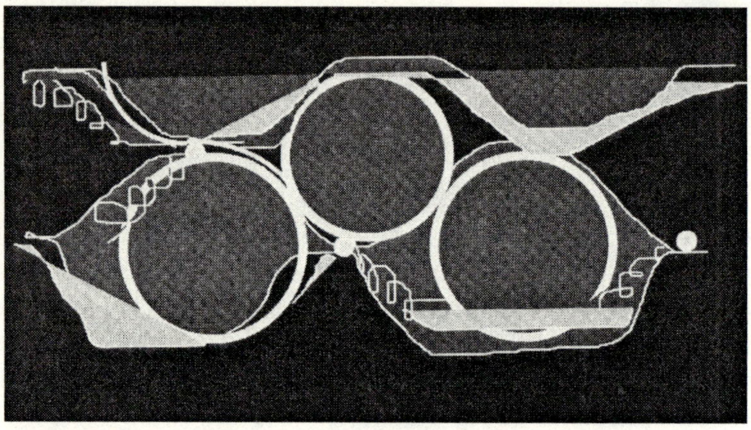

Above is a partial pattern of cycles. Overwriting it are force formation correlations. Yellow dots represent the macro spherical spinning formation of the strong force, or big bangs. And they're at our cyclical junctions. These then branch out into light blue spirals. The dark blue circles represent a gravitational formation through the massive nonlinear scale of cycles. In reaching the nonlinear perimeters, the gravitational formations describe massive expansion and contraction stages. And they also seem to amount to the cyclical delineation per se. Don't ask me where the macro waves are.

Speed c theoretically, is a nonlinear demarcation of mass-motion fusion in spacetime. And you don't get it on the linear plane. Your wave, however, depicts the speed c formation. We're recognising the animation of spacetime through nonlinear motion using these formations. That's how I'm gauging energy and mass transformations. A combination of motion, space and time manifests a combination of energy and mass. Well, you may think mixing up motion and mass this way is putting chaos back into the equations. Not for me. Who's doing equations anyway? The thing is; these theoretical force formations derive through a structural basis of symmetry. Any animation and combination of spacetime and motion should remain applicable to the balance it provides. It means superluminal activity is not necessarily chaos. Symmetrically, we might decipher meaningful energy and mass transformations *beyond* speed c. In other words we're looking at phenomena through a different angle. There's no black and white about this anymore. We're watching the balancing act of all the colours of the rainbow.

Reasoning phenomena through a spectrum where motion is inherent within spacetime, shows transformations arise through *curvatures of motion*. Motion is curving along with spacetime, which I think is a mind shift away from the concept of velocity. Motion you see, should be as malleable as space and time. Indeed motion could take on the appearance of encapsulated or bonded mass. We're not judging a body in motion with this. Go to gravity for that. Rather, the nonlinear aspect fuses the mass and motion somewhat. And the point is, we could be reasoning mass, not just energy, as a form of motion. $E=mc2$. Come on, wake-up brain. It's not that difficult. I want to try re-arranging it later. (The equation and my brain.) Already we seem to detect mass-motion on the micro level with our quantum uncertainty and the wave/particle duality. I'm just suggesting that could also apply as motion condensing into spacetime such that it figures as solid motion. And sure, I know it sounds ludicrous. Mass is motion. Motion is mass. We're going superluminal with it. Believe it or not. Work it out or stay in the bubble—because right or wrong, I'm giving it a go.

For example, the spherical spinning formation might represent motion enclosed and bonded through space and time into a form of mass. The charge of the nuclear proton amounts to something approximately 2000 times heavier than the electron, apparently. Now is this because the proton's charge formation comprises of 2000 times more confined motion than the electron? Consider also; the space that the negatively charged electron moves around in, compared to the space for positive charge within the spherical spinning formation. This nonlinear concentration of motion through a spinning spherical formation may be the nuclear binding energy of mass. Moreover this formation, involving mass and charge, shows its force declining over increased space. It seems that the lines of force are subject to the confines of motion, and its condensation into mass. And I think we're very superluminal at this stage.

For symmetrical purposes, space and time merge with motion in four ways. I've said it before, but it's important here. So each force formation should have a different effect. Theoretically they transform matter and energy through their unique mixture of spacetime and motion to realise *nature's diversity*. Well, this diversity could be an ongoing occurrence according to the symmetry of the system that the four add up to. And all of that could be subject to the greater systems within eternal and infinite spacetime and motion. But let's not get too far ahead of the point. Whatever it is, we'll never get to it digressing into eternity.

I think symmetry imposes an order within spacetime and motion that conserves it according to energy and mass transformations. Now I don't think you could arrive at this symmetrical consideration without motion in the spacetime rationale. Our four force formations are all powerful to this end in different ways and measures that, in some sense, are parallel. Consider the power to unravel nuclear spherical spin compared to winding it up. Could this form of power, that of the time and space—as well as the motion necessary to condense into a nuclear spherical spin, accord with a nonlinear reasoning of gravity? One that correlates with gravitational expansion and contraction? Is this an example of the power of symmetry? The effect of one disbursement of space, time and motion, is perhaps producing a balance through the formation of another. One, that is, with a different combination of spacetime and motion.

Motion can give a different meaning to spacetime. Evidently it empowers it. The point however is this; spacetime also empowers motion. Condensing it down is giving motion loads of impetus. It's empowering motion into mass.n b1 But reams of spacetime scarcely infused with motion strikes me as pretty powerful stuff as well. Using formations you can liken the total power of one force with another through symmetry. One with maximum spacetime and minimal motion might amount to one with maximum motion and minimal spacetime. So they function differently. Yet they should all amount to the symmetry of the same system. Where gravity has an abundance of one element of spacetime and motion, the nuclear forces have a great deal of the other. If space, time *and motion* are all subject to a symmetrical distribution, these force formations should have different amounts of each to balance and conserve the ongoing totality of it all. And their different abilities to transform mass and energy should all lead to the conservation of everything.

Already I'm suggesting the mass-motion mixture, or fusion, accords with the lines of force of the formation. The more it fuses the less far reaching the force is. The power of the strong force, for example, has more nonlinear mass, or condensed motion. As such it accords with reduced spacetime. Its lines of force tightly knit into a small spherical ball of spin. And here we go again with the wave. Compared to the strong force formation, the wave mass-motion mixture is far more diluted. I think motion is just about on par with space and time here. It's got more space than spherical spin, producing lines of force that are effective over a greater area. And as it's still nonlinear, (the cutting edge, remember) the formation should still be able to disburse motion *through* mass/energy exchanges. Again, we seem to have arrived at the wave/particle dual nature of light.

Now back to the systems logic. How's your head, by the way? I realise there's a fair slab of reiteration here. And we're not shooting the breeze of four dimensional spacetime with this logic. We're going through four nonlinear formations. And they systematically function on increasingly larger dimensions. Spheroid simplicity, it isn't. Well, it can be. Think in dynamic spacetime. Generally things go up and down, in and out. We're just animating that four dimensional framework. Starting with the intense nonlinearity of the spherical spinning formation, mass and motion could transform and stabilise through spiral, wave and orbital formations. And that's on different levels. You've got your atomic, stellar, and galactic systems all showing matter and energy transforming though these formations. Reality speaks for itself. In all of those levels, these formations could dilute the intense nonlinear curvature of spacetime and motion until mass seems to exist in motion rather than as a mixture of it. Then once again, you can think up and down, and in and out with it.

Take photons. With zero rest mass, photons of light present a mass-motion mixture travelling within the wave formation. Well, that business could realise by way of a definitive micro system. One that comprises of four force formations in all. You see; corresponding to the energy and mass states of atom transitions, are the distinct wavelengths in which they travel. And through the wave formation, photons can transfer mass-motion rather than mass *in* motion. Taking excess spin out of the atom thus, maintains atomic stability. The system stays symmetrical. We've got spirals, waves, orbitals and a spinning sphere all in this together. I've already gone through it. Your orbital levels of electrons, through which the photon exchange occurs, appear to clarify the particle formation of mass from nonlinear motion. And the orbital formation has the lowest degree of nonlinearity. It's the least dynamic, and where you can size-up four dimensional spacetime. But here in the micro set-up, you don't seem to get a predominating affect of gravity with the orbital. It seems more concurrent with the system's other three stages of nonlinear intensity. Perhaps here you can also decipher the symmetry of the system per se effecting energy and mass transitions.

Remember though; the orbital formation should realise mass with minimal inherent energy and nonlinear motion. This is because it has the greatest ratio of spacetime to motion. And that applies on any systematic scale. Indeed, the mass of gravitational attraction always appears to be linear, only functioning below speed c. Nor does it seem capable of reverberating into intense nonlinear mass-motion. I mean your orbital stuff shouldn't directly become your spherical spinning stuff. Or, could it? What seems stable today is a black hole tomorrow? It depends on the bifurcating level of the system.

Look at the cyclical level. A macro slab of sub-systems can condense into the nonlinear intensity of nuclear spin through this framework. I'm talking about a cycle of spacetime and motion. Everything in one could *radically* convert back into an intense nonlinear mixture of mass and motion. There's your pin-point of spacetime and motion. Here we have the large scale destabilisation process of everything through a 'big bang'. Or call it the general relativity of the strong nuclear force. On this level, the nonlinear dimension of a gravitational orbit shows a spatial dilation of mass. After a certain time at the same velocity level, it can eventually produce its nemesis. And in the meantime greater spacetime and motion maintain a symmetrical balance.

That doesn't seem to happen in the atomic system. There, your force formation interactions apparently progress along the scale of next in the motional line. Taking into consideration their concurrent symbiosis, that is. Your orbit wouldn't dive into the spherical spin. But on the grand scale cyclical level it looks that way. Read on though. We're not there yet. This issue is for later chapters. We're going into black holes, quasars and accretion discs then. I'm

hoping these things will describe a more stable sequence of formation events. One thing though; it's becoming increasingly apparent that you don't go superluminal with some sideline analysis.
We're covering the whole jolly cosmos to get there.

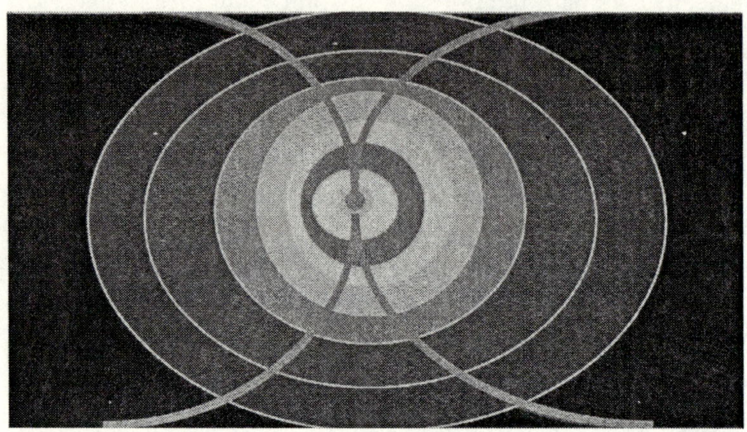

In the above diagram, partial pink orbital formations outline gravitational contraction and expansion. This is our macro level of cycles.
I reckon here is our limit to symmetrical invariance. We're down
to one formation deciding a whole framework. And we need four
for symmetrical validity. With four you've probably got your plural set-up. Anyway, this diagram shows limitations and links between four cycles, using two formations. So where's the big spiral
and the big wave, you may ask. Where indeed. I figure; when they
predominate, cyclically, we're in the quasar and the accretion disc.
And at this stage I'm not drawing that. This is about the all time
condensation of mass and motion. For brevity I'm just outlining
the orbital formation and the spherical spin here. In the drawing,
the completion of one formation kicks off the formation of another. A cyclical 'end-point', for instance, could realise a spherical
spinning formation. See the small pink circle. There's your limitation of massive contraction. For balance, a new cycle begins through
another orbital expansion. It should be inevitable anyway. With
motion in spacetime, symmetrical balance is never static.

The thing is, mass, theoretically, isn't absolute. It always ties in with motion to some extent. We just have to reach the cyclical framework apparently, before we can see that's the nonlinear case with gravity's mass. Cyclically, you get the full curvature of the orbital formation transpiring into another formation. That's when our nonlinear dimension heightens. Compressing spacetime with motion at the cyclical junction fuses motion into the big bang mass situation. A pre-big bang scenario, perhaps? Would it be so absurd? Tell me about everything coming from nothing instead. That's your popular alternative. At least this way it derives through symmetrical necessity. According to this theory everything's relative. Time, space, motion, energy and mass all bear some relation to each other. *All the time,* that is. Nothing is absolute. Everything is relative. And it's always moving. So where's the cosmic corner for drawing everything out of nothing?

There's nothing complex about that. Here's a simple story of the cosmos. Sure, it's no magic number, but maybe some exist that apply to it. We've got four force formations of spacetime and motion. To preserve the eternal and infinite nature of the cosmos, they interact according to a symmetrical disbursement. On that basis, mass and energy remain subject to an order—a symmetrical structure. Where, for instance, the gravitational field is strong enough, nuclear matter might well compress. I'm not denying it. But going by the above, that's only up to a certain point. It never compresses into an absolute situation. You can see it on this grand scale by the symmetry of your formations and their spherical spinning pin-point junctions.

You've always got greater forces of symmetry influencing a situation. When it's gearing up towards an infinity of totality they step in. So your gravitational/nuclear compression would transform accordingly. And that seems to be when the cyclical transition occurs. Besides, space is never out of the equation. Motion and time don't function without it. And when we talk about infinities

I think we're reasoning stuff outside that premise. You shouldn't get a worm hole out of a black hole this way. Indeed, you wouldn't get hole at all. Your black hole is still a form of spacetime and motion. It's just an extreme combination of them. We've also got an ordered external spacetime and motion happening. It can systematically react to internal dynamics. That means if the nonlinear situation becomes sufficiently intense it registers an imbalance within the system. Nuclear matter could transform into the mass that eventually generates a gravitational field. ('Eventually' being the key word.) Theoretically however, the imbalance should initially realise a spiralised reaction. And here on the macro level, these formations don't figure in tandem. They individually predominate. Either way, all things considered; the compression would ultimately trigger off a new cycle. Then the whole bifurcating set-up supposedly happens again.

I think therefore, the invariant nature of symmetry, through cyclical links, overrules infinite density. Symmetry defines mass and motion and how far we can stretch them. Your force formations' limitations structure it. The point is; reasoning through them could help us work out mass beyond speed c. Look, both micro and macro systems of force formations involve symmetrical necessities. You've got your charge release of spherical spin imbalance on the sub-atomic level. Then there's the nonlinear ramifications of gravitational orbitals on the greater scale. On either level the system progresses beyond the speed c barrier. We've got spirals and spheres after the waves. All we need do is untangle the energy/mass mess of the nonlinear fusion these formations manifest.

For this reason I'm going into Einstein's equation. Why coin a phrase without it, anyway? Symmetrical bifurcation can mean bugger all. Try it on someone. All I get are vacant stares. And there's no simple explanation. So why bother? But bandy a few numbers around . . . even statistical misnomers probably register something. I guess they're like tools of reality. Thus far I've shoved them off the work bench. Well, not anymore, baby. Numbers are big time and can speak volumes. In the next section I'm giving them a go. I'm going to dance around their totem pole and outline hypothetical maybes. So what did you expect? In depth mathematical cogence explaining superluminal activity and the cosmos in a numerical nut-shell? You must be joking. First I'm going to study the form and practice on the pokies. That's 5 credits on 1 line, or 1 credit on 5 lines using a 2 cent machine. Or we can even go 20 credits on 25 lines. Now with your diamond machine it amounts to 150,000 credits if you crack it. Numbers land here we come.

70 bucks down this morning. So much for practicing on the pokies. Clearly the pokie is programmed 75 to 25 in favour of the club.

. force formations and E=mc2

This is wishful thinking, you know. Maybe superluminal travel is just a high ideal, but I reckon we can reach it through reality. Translating it into equations however, is total fantasia. There's more chance of winning Lotto than inadvertently stumbling across the elegant brilliance of the cosmic number by yours truly. I couldn't numerically say what a spiral amounts to. Give or take a bit of repetitive rhetoric, however—expletives edited out, we might be able to make some sense of them according to E=mc2. That brilliant equation is my bridge over the murky superluminal mixture of energy and mass. You see; smack bang in black and white we've got c squared here. There it is, with energy and mass. And it's proven beyond the back of Bourke. It's proven as far as the speed of light can take us. I'm just going to see if it can take us right up to its outer limits. That is; the c2 of energy and mass. So I'm sticking it on the mental dash board and here we go. And mind, it's a third-hand Ford F250 we're travelling in. It's got a V8 though with Tassie plates.

First I want to boil down the relationships of motion, space and time with the energy and mass transformation processes. From the above, it seems a higher degree of nonlinearity in the force formation means more mass-motion fusion. Well, I'm wondering how that accounts for energy in the E=mc2 equation. Theoretically, energy and matter ratios are also subject to the curvature of spacetime and motion. Remember, these four formations are the symmetrical differentiation of one force of spacetime and motion. They all amount to each other. So where one has more space it also has less motion and time, and so forth. Ditto then; mass and energy. And mass and energy transpire through the forces of spacetime and motion.

Now the spherical shape of the nuclear force seems to involve minimal spacetime to maximum motion. Say, 95 parts' motion to 5 of spacetime. On the other hand, the curvature of a gravitational orbital has a great deal more spacetime, and relatively little motion. Maybe that's about 10 to 90, or even 3 to 97, favouring spacetime. But the thing is, gravity seems to involve matter more than energy—whereas the nuclear forces; energy rather than matter.

On that basis then, the ratio of energy to mass in the subsequent spacetime motion transformations, depends on how nonlinear the formation is. Using the mass/energy transformation equation we might be able to see at a glance the mixture of space, time and motion. We could even end up gauging the symmetrical impact. For example, ec-= *mc+ would concern something more nonlinear than a wave. We've got motion above speed c in the bit* pertaining to mass here. Theoretically, that describes a spiral of weak nuclear force.

The relativity of matter and energy should reflect the relativity between space, time *and* motion. In some form or other, matter and energy derive through the eternal transformation of spacetime and motion. And I don't think the quantum zoo covers all that. Nor can general relativity and quantum physics through their disparities. Ideally, the conservation of matter and energy accords with the eternal and infinite nature of the cosmos through symmetry. Realistically, it probably does too. Theoretically it does—check out the force formations. Numerically, we're whistling in the wind with it, according to symmetry. (Or is that the other way round?) Even so, the equation E=mc2 recognises the law of conservation. Matter and energy always relate through this proven equation. It just seems that when we reach speed c, matter and energy don't figure through spacetime and motion. And this is despite c2 factoring within the equation. The nonlinear theory, however, broadens the context. We've got ongoing transformation processes for starters. Motion necessarily in the spacetime concept affords this. Using these force formations speed c motion is a symmetrical application rather than a full stop. And it figures in both the force formation as well as the energy/mass equation.

With the E=mc2 equation as a framework, I want to draw divisions that apply to each force formation. I want to map out the symmetrical footprints in the sands of transformation. Hypothetically speaking then, and on that basis, the motion maximum within the spacetime concept is c2. I'm going to stagger it alongside spacetime relative to energy and matter transfers. That means four levels of motion up to speed c squared apply to four combinations of energy and mass. Clearly it's not so simple. There's countless zillions of combinations of mass and energy. But I still think we can come up with some broad guidelines.

Waves, for instance, travel at speed c. Their energy and mass content could be equal according to speed c. I reckon with the wave we're also looking at equal portions of spacetime and motion. Indeed, I think ec=mc would be the symmetrical wave equation. Energy at speed c equals mass at speed c. That's the wave. O.K. then, it's a hypothetical one of tsunami proportions, sure. What's energy at speed c? Is it hot, cold, blue or white, or what? I don't know. All I know is that it takes a wave formation. And it bears a symmetrical relationship to mass. Look; the equation is also describing the ratio of inherent mass-motion fusion. That's our degree of nonlinearity. So you get an idea of the formation. Your speed c mass-motion fusion is contorting spacetime to the wave format. The wave, or ec=mc, disperses energy and matter evenly at light speed. And pin-point your mass position in it with anything less intricate than this wave dispersion of energy and mass, you'll lose your velocity reading—and vice versa. Call it symmetry in motion. And try a spiral instead. I mean we could probably probe the wave with the spiral.

Where once we would calculate impossible mass increases at light speed, we can now calculate its symmetrical transition into energy and nonlinear mass-motion fusion beyond speed c. Your impossible mass increase becomes a spiralised combination of motion instead. It takes on a different form. It's subject to a different force formation. It fuses into the motion of spin. And it can transform through a spiral of spacetime. (The spiral has some spin in it.)

I'm going through it again. Even equations at this point probably wouldn't simplify the line of logic. The force formations describe a *formation* of motion throughout spacetime. It's a decisive factor in transformation processes of mass and energy. The formation is the symmetrical imposition. And we've got four of them. They *affect and effect* energy and mass transformations. And later on down the line, it seems that energy and mass do that to them. Take macro gravity, for instance. Anyway, translating the force formation into the mass/energy equation means recognising the formation of motion within transformations. For example, registering anything beyond c in the mass side of the equation could be recognising spin. There's the mass-motion fusion of intense nonlinearity involving a formation of spin. Indeed, spherical spin is theoretically the mc2 of energy. You've squashed spacetime into a completely nonlinear nuclear nut-shell. Which is thus rendering motion proportionally astronomical? Look; we've spun it up. We've intensified its dynamical situation. We've upped the motion factor. And we've impregnated spacetime indelibly with spin. Are you with me? It all translates through the system. Subsequent energy and mass transformations subject to this force formation should all bear some degree of spin.

Thinking about it, there's some spin in everything. The planet, the electron, the proton, my brain—all have it. I guess how much spin shows up in the transformed mass, could depend on how closely the force formation works in tandem with the rest of its system. Your electron spin, for example, could carry more spin a la the proton than the planet does via the sun. If so, our force formation equations should also recognise, and reflect, the system per se, to some extent. And so it should. It's symmetrical validity does to an indirect extent. All four formations are symmetrical derivatives of each other. You should be able to see more than tangential relevance between them. I mean they define each other. And on your sub-atomic level that seems to happen practically simultaneously. So your energy and mass transformations could be contingent upon the dynamics of the system's symmetry generically. I know . . . I've said it before.

And that's not all. We've still got the bifurcating business. One system is supposedly subject to a greater one. For example, strong force motion could amount to 30 million years. That's supposedly how much time and space we're spinning up. Spiral spin would be considerably less. It's strung out more. Well, here I'm gauging the spacetime of proton spin according to the symmetry of a stellar system, rather than its own atomic one. I'm going by the average central proton in the sun. You're apparently looking at 30 million years before one such proton unravels into a neutron, neutrino and positron. Our ancient proton could be unravelling according to the greater symmetrical scheme of things.

On the other hand, bugger the sun. Isolate a proton's atomic set-up and it might not take 30 million years to unravel. Look at your atomic charge set-up. Your proton dynamics per se could set off its own symmetrical imbalance. Well, at least it might happen less indirectly—as per the greater scheme of things breathing down its spinning neck. Remember; all that theoretical stuff on symmetrical imbalance propagating weak force spiral activity? Remember; it realises electron orbitals and radioaction beyond the atomic shell. I'm wondering whether that only derives through the greater symmetrical system in the bifurcating logic, or whether we can get the ball rolling without it. If we could up-end our proton—strum up some imbalance—would it spew forth superluminal power? I think so. The trick is to control the energy/mass transformations in the process. You see; the spherical spin curvature is not simply maximising motion. It's also fusing it into mass.

The real trick, however, is to do the equations. Evidently, getting one out of me thus far, is tantamount to Sooty reciting the Man from Snow River in Sanskrit. Witness the procrastinations, reiterations and digressions on the way. So much for talking numbers.

But in whatever language, unravelling the curvature of spacetime and motion of a force formation, theoretically, should realise a combination of energy and matter according to the formation's level of motion. At some stage, say with the nuclear spherical formation—and according to E=mc2, we should release the energy amounting to motion of speed c squared. Nothing to it. The thing is, I don't think you can forget the bifurcating business. Each system of force formations could have an exponential value according to the greater system surrounding it. The more micro, the more these formations proliferate. And the more concurrently they interact with each other, etc.

Just imagine putting all of that into numbers. We've still got the time factor to consider when we unravel the spin, as well. That means factoring in the bifurcating imposition. And don't forget the symmetrical translation of speed c. I'm putting a figure on the curvature of motion here, not just spacetime. Then we're *ostensibly* calculating energy and mass transitions according to it. Well, I'm not doing that. And for that matter who could? See what I mean—wishful thinking. Your nonlinear equation is a mind bender I believe, that even Einstein might've found challenging. He might've found it so challenging that we never progressed from its mind boggling provocation.

In lieu of nonlinear equations then, my challenge is to metaphorically figure it all out. Failing that, I'll try and be linguistically accurate. How do these force formations convey mass that's not a nonlinear mixture of energy and motion, up to and beyond speed c? Let's return to the sub-atomic spherical spinning formation. Here's a condensation of nonlinear motion that looks almost solid. Indeed, this mass-motion fusion could show up as the consequent conveyor of the force, rather than the engine of energy and mass transitions. As such, we might be missing the spin factor of mass-motion fusion transforming down the nonlinear line. We might not see that factoring through other force formations and their interactions into matter stability. We might miss the symmetrical links and its translation into energy and mass. Remember; according to our $E=mc2$ logic, transformations directly pertaining to this formation should involve more energy than mass/matter. n b2 We don't get to mass stability until we arrive at the gravitational orbital below speed c. And that's when spin is all but indecipherable. Well, it's no longer your major measuring device.

Sub-atomic particles seem more enmeshed within the force formation. There could be underlying patterns and relationships among particles and forces here. Even so, I think we've got a lot of intricate nuances confusing the clarity between energy and mass. Stuff that is, besides my maverick thinking patterns. Quarks, for instance, might be ephemeral mixtures of spacetime and spin amidst transformation. That's rather than clear-cut combinations of energy and matter *in* spacetime and motion. To date, our forces relate to such particles. I'm suggesting these particles are first stage energy and mass transitions still pertaining to a common underlying symmetrical application.

I think the force itself is the engine of transformation. It's the symmetrical differentiation of spacetime and motion. It thus carries a formation that realises particular energy and mass transformations. In this way our sub-atomic particle would be an outcome of force interactions, and indeed an expression of them, rather than the conveyor of the force. Your photon, for instance, expresses energy and mass transitions subject to the wave format. But it also derives its fuel and form from the spherical spin and orbital. (And I reckon spirals figure in there too somewhere.)

It always seems to come back to spherical spin. And without the orbital, spiral and wave, perhaps spherical fired spin might not unravel into mass and energy. It could come out as full on motion. Isolate the proton though, and don't we get neutron—neutrino, et. al. transformations? Well, more of that later. I'm pre-empting the conclusion.

Now at the other end of the equation framework, there's the orbital formation. Here the combination of spacetime and motion is very different from spirals and spherical spin. At the orbital level the nonlinear confusion of mass-motion seems to have levelled out to render the stability of mass in motion. Motion, thus, untangles itself from nonlinear intensity to below speed c. Lets' say we're taking the spin out of it. In other words, energy divided by c2 arrives at gravitational mass. And more mass, as such, means more gravity.

You don't seem to confuse transformations of mass and energy within the orbital formation like the other formations. And there's no decipherable mass-motion links of nonlinearity. For example, gravity bends the trajectory of the moon into an orbit around the earth. We don't see the moon transforming in and out of energy within the orbital formation. At least we can clearly see that the moon is not a conveyor of the force but an effect of it. Here we're getting the linear clarity of maximum spacetime defining mass from motion.

But let's go back inside the atom. Our sub-atomic orbital is not as clear cut as our lunar one. Even so, there's still heaps of comparative space in between orbital and proton. I mean the maximum spacetime quota to minimal motion, pertaining to the orbital, applies likewise here. However, our energy and mass transitions look more susceptible to the concurrency of the system. When you locate mass stability in an orbital, you're still looking at wave formation and electron spiral transitions.

In summary then, I think we can lay down a few basic equations here. Well O.K., they're not equations exactly. I'm making some general deductions. The spacetime and motion percentages of each force formation appear the same on any level. Further, mass and energy transformations relevant to them would reflect that. The depth of symmetry translating through them into energy and mass however, depends on the system. We don't seem to arrive at the stability of mass until we're in the orbital level. And even that isn't clear-cut until we transcend the atom. So our sub-atomic energy and mass equations should still reflect the spin of speed c and it's mass-motion fusion to some extent, next to each force formation.

On any level however, energy and mass transformations concern the *system* of four force formations. That's a consideration for our hypothetical equations of energy and mass relating to each formation. They should reflect transitional stages en route to mass stability. In other words, all energy and mass transitions are bound up within a system. Look at the stellar system. The spherical spinning set-up of the sun hardly resembles mass in stability as the planets would. The planetary orbital seems to demystify the last traces of mass-motion fusion. Per se, it's a different story, sure. We've got our bifurcating logic to consider once more.

The total mass of the stellar system should effect a gravitational force according to a larger set-up. Say, a galactic one. And as we go up the bifurcating ladder, our force formations don't happen so much in tandem. Nor do they proliferate so much. Let's say they tend to go in some sort of chronological order rather than full steam ahead all in line. And your systematic display of mass and energy reflects it. We get a spiral galaxy following some other major display. But on the smaller stellar level of greater bifurcation, we don't seem to get a total spiral set-up. For that matter, I don't know how the spiral figures in the stellar set-up. Theoretically, it's in there somewhere though—along with wave activity. The thing is, here they should figure more systematically than on the galactic level, but less concurrently then they would in the atomic one. In other words, where the orbital is happening in the stellar system, the spiral should be less evident. In the atomic one though, you could have orbital and spiral forming and happening together. Whilst on the galactic level, it seems to be one formation at a time.

Finally, without the transformation of mass and energy, forces seem meaningless. The two, apparently, instigate and support each other's existence. Indeed, force formations appear to rebound through the symmetrical expansion and contraction of mass and energy within spacetime and motion. That's your systematic bifurcating logic again. Your greater systems could manifest and function subject to the energy and mass transformations of sub-systems. Or is that the other way round? On the macro level, mass seems to generate gravity. But in your atomic system, the orbital corridor of spacetime makes way for mass in motion. And on the cyclical level, the strength of gravitation attraction can render the whole mass and gravity equation into a big bang of spherical spin once more. The point is; the bifurcating set-up could hold clues to those impasses between the laws of gravity and nuclear forces. Theoretically, you can't just compare, willy nilly, the energy and mass manifestations of one force to another. You have to first recognise its symmetrical position within a system of four forces. Because, you might be comparing a force from one system to that within another.

When we get to our cyclical level I think we've played the game out. Our force formations and energy and mass transformations systematically feed off each other until they're at a symmetrical impasse. They can go no further. Finally, our grand scale gravitational orbital is also subject to next-door's spherical big bang, etc. In other words, our systems are merging. One cycle defines the perimeters of another. One's orbital expansion is also another's orbital contraction. Remember, on this scale, one cycle's mass ratio would not directly activate the systematic *predominance* of another force formation. Now it depends on the symmetry of a cyclical framework. Your system is subject to the greater cyclical setup. And we're already picking it up indirectly, perhaps.

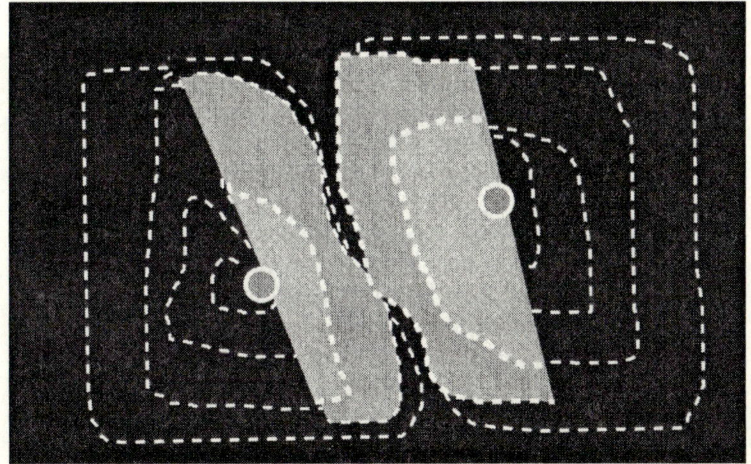

The above diagram shows my eyes at this stage of the game. Or is that the segmentation of two adjoining cycles? They both have an internal structure of decisive mass transformation levels. An expansion of one could affect the contraction of another.

. endnotes

1 nb mass is different from matter
2 nb no doubt the mass/matter distinctions here are important,
but I'm not getting into them

ch.7

a context for superluminal thinking

. *preamble*

Why do we think as we do? For starters, we've got general thinking patterns. I know we're all subject to capitalism and other lateral trends. We've got free market forces defining the ozone layer. We're self declared independents to the point of licentious ennui. Well, a large whack of the population is. What's this now—the age of the orgasm? I'm wondering though, is there a framework beyond our lateral thinking patterns, underlying our concepts of the cosmos? It's a subject that pretty well runs through out the text. In the beginning. . . .and then you can burn in hell forever or live in eternal salvation. What's with the beginning? Why are they still there in the rationale? We probably won't see beyond speed c with them. We could end up stuck in our bubble until it bursts on us rather than metamorphose out of it. That's our conceivable reality. And where's the logic of eternal salvation or damnation? We've got 2 extremes of the same coin here. Alongside a definitive genesis or a big bang beginning, *eternal* doesn't make sense.

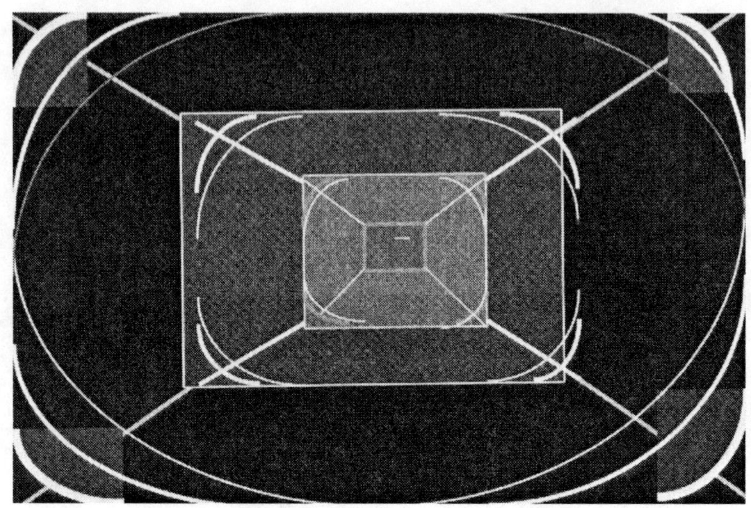

Take the spacetime concept, for instance, where space begins with time. This applies to both general relativity and quantum theory. In fact I'm taking it as the standard bottom line for cosmological enquiry. n b1 On the big scale, and inside the atom, we're calculating stuff with a finite attitude. Nature in all respects, for us, has a beginning. And motion doesn't necessarily factor within its spacetime. We're not getting the dynamic dimension. I'm not denying the realities of the concept, though. I'm just saying the concept doesn't cover everything. Inconsistencies appear at the extremes of both the macro and micro level.

On the micro level are principles of uncertainty. Whilst at the other end of the spectrum, we can end up in a black hole without an exit. They're waiting for us at the outer edges of general relativity. And on any level, we just keep expanding for no particular reason. It doesn't ring true. The greater reality of the cosmos is clearly still eluding us. Anyone can figure that one out. We're frigging around in the bubble still trying to line up the gravitational pull of the Milky Way with the spherical spin of a single atom, more or less. Or so it seems. In working out the starting point for one system, and that of another, have we missed the continuity of each and connections between both? Look, I know it's only the conjecture of a self-informed hill-billy. I could be wrong. But then again, I could be right.

Anyway, for what it's worth I'm questioning the paradigmatical structures these theories employ. I reckon they're the same. There could be corridors in our thinking space that are shunting us into pockets of reality. Once in them, we can head down their rapids and up their glaciers, but you don't get the overview of the valleys and mountains together. Who can see through the sun-rises of Seyfert galaxies, and the sun-sets in the Coma Cluster—Hubble's descendants in the year 5 million and ten? You couldn't with a sun beam. You couldn't go the distance in one lifetime. And besides, you can only see as far as it lets you. It's defining our bubble. It's only giving us the twinkles of the other billions—tantalising twinkles, that keep you wondering.

. paradigms

> north is probably north of north,
> and south is probably south of south
> but is that linear or what?

What's a paradigm, and why is it so powerful? I'd never heard of them until university. And then, not wanting to look less erudite than the rest, I never asked. Nor did the dictionary help. Paradigms then, could possibly mean something off centre for me. In this context, they come over as a metaphysical cosmic framework. One that is, that outlines the spacetime and motion for thinking. You can only grasp the cosmos in so far as your paradigm allows you too. They seem to impact significantly when it comes to connecting theory to reality using the scientific method. I think that gives it a psychological power as well. Once a paradigm prevails, any experience and observation is explicable according to its terms. It doesn't just provide a structure, or working model, through which science can act. I reckon it also applies to any belief system underpinning your theory.

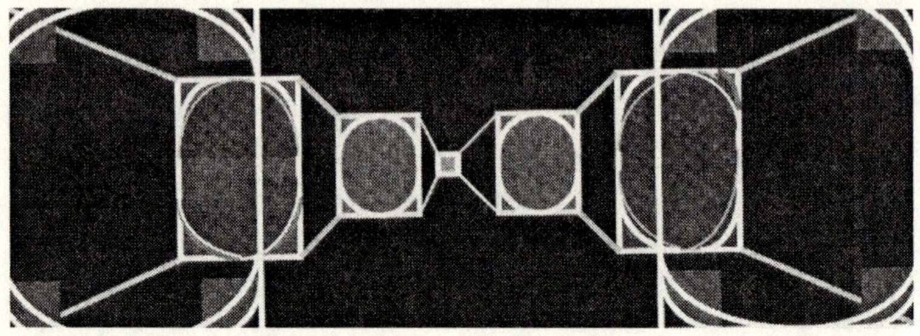

I guess when science fiction subsides to a dearth, your paradigm is past its used by date. Strange as it might sound, that could signify the inadequacy of the paradigm. Reality could be overwhelming it. And the reasoning could become retrospective. My concern, however, is that the paradigm might also produce tunnel vision. Your cosmic rationale is no doubt subject to the focus of the proven paradigm of the day. Take the paradigm of a finite cosmos. (Take it away.) This paradigm might direct observation and experiment into conclusions according to a bubble. Phenomenon—conceivably all of it—figures through a starting point. It probably begins with a big bang and then expands spherically outwards. And one thing leads to another. Given time, the emphasis goes on finding 'initial conditions' data through scientific experimentation. In this way, the paradigm might 'prove' reality according to its boundaries. Where the boundaries are easily accessible it wouldn't be that difficult. You wouldn't have to think too deeply about alternative scenarios, would you? I mean you wouldn't have to tap into those creative corners in the nether lands of your sub-conscious and come up with some interesting options.

But who's going to question a paradigm with overwhelming data supporting it? You're up against the predominant way of thinking about the cosmos. That's what it realises. Any new theories are subject to the one that fits the facts to date. Besides, today the paradigm presents a working model for stuff, theoretical or otherwise, in a world attuned already to realism and pragmatism. It won't just outline and control the borders of what's up for scientific reasoning. It's also defining the scope for imagining and believing. In readily reinforcing the reality of the day, it can thwart new supposition over paradigmatical borderlines. You've got to reach into the unknown for that. And your prevailing paradigm's discrediting such wonders with its all encompassing validity.

And try questioning reality beyond the paradigm. You don't ask, what happened before the big bang, or decide that quantum uncertainty is bullshit. Well, O.K. then so it's there, we've proved it. But doesn't that also mean a failing on our behalf to reason through it? You don't conduct experiments on the basis that the universe is infinite. And you don't describe the results of an experiment along these lines. Most cosmologists apparently reject the notion of infinity because of the second law of thermodynamics, anyway. But is this law describing a closed system? One that is subject to the finite limitations in spacetime and motion of a prevailing paradigm?

So here goes nothing I suppose. We're going to look over the fence of reality. I want to transcend the prevailing paradigmatical borderlines. We might as well go for the biggest paradigm possible then. That is, one of an infinite and eternal cosmos. I'm opening my mind to it. Then I'm drawing up some defining perimeters. And there's the challenge. How can an infinite and eternal cosmos have perimeters? Well, we've still got the marginalised stuff of the prevailing paradigm to kick us off. There's all that observation, experiment and its extrapolation. I'm bearing in mind though, that this stuff is below speed c. We've been crawling around various areas of the cosmos at a snail's pace, only catching twinkles and glimpses of pocket realities. And through our grid of reality, their radiant below speed c glow only turns into the brilliance of cosmic expansion.

A working model of eternal infinity should go beyond the ballooning set-up, although it should still include it. As such, we can take the ballooning borderline into a cycle. In various stages of expansion and contraction, we can reason many balloons. They could all link in with each other across endless spacetime and motion. That being the case, our prevailing paradigm's borderlines should come over as cyclical changes. Previous limitations may show up as structural mechanisms. You could see the reason of symmetry instead of finite dead-ends. Your bigger paradigm could interpret data, such as that of a big bang, as evidence of cyclical changes.

In our reality though, the prevailing paradigm appears inviolable. We ignore or ridicule anything new and inexplicable according to it. Where data doesn't fit in with the paradigm, it's the data rather than the paradigm that gets debunked. The thing is; stuff is building up at the boundaries of the laws of physics. You'd think therefore, that the laws of physics would threaten and challenge the paradigm's boundaries. But they don't. They work so well with the stuff in between—in the bubble. And so we're not coming up with any new laws of physics, are we, after all. All we're getting is worm holes and time reversal scenarios at the edges of our thinking space. That's why I'm suggesting we initially question our paradigm's borderlines by way of a bigger set-up. Then through the extenuating circumstances, we might deduce some greater laws.

I wonder how many discoveries we miss out on because of the established mode of thinking. New stuff continually shows up that doesn't fit in and still the paradigm's borders appear insurmountable. So much for overwhelming them. Rather, the spacetime paradigm looks most active when we confront the limits of knowledge. Take particle accelerator experiments. By reaching to the edge of the paradigm they don't prove everything. Where's the four force unification? Apparently you'd need 10x28 electron volts and a collider 1000 light years in diameter to work it out—could be a cyclical perimeter but who's seeing it that way with the present paradigm? So you stay with the below speed c realities that restrict experimentation accordingly.

Forgetting science for a while, what happens to society with an actively inadequate paradigm? Thinking could become introverted and circumspect. Reality may seem regressive and the arrow of time more questionable than ever. Your future could press in on the present without the scope for a guiding system to evolve. I mean your present could become a foregone conclusion devoid of depth and meaning. Your future could be a boring repetition of that. How does society function if it can't develop and spread its wings? Is this when cynicism, immorality and broken dreams manifest? Is this when our realities become a retrogression, rather than an evolution of spacetime and motion?

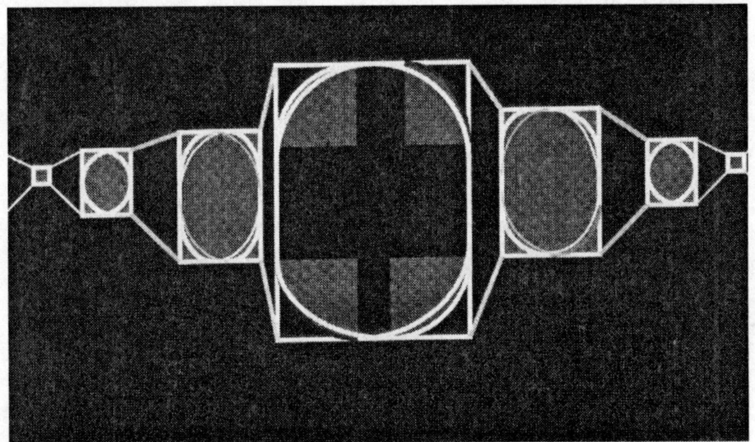

I think it's happening already. Check out the latest fashions and music scene. Check out rising youth suicide and heroin addiction. Witness an era devoid of mythical heroes and heroines. Look at the art. I don't anymore. What's inspirationally uplifting about someone's nightmare imploding on a wall? Where's the beauty? You'd think with all the orgasms going round we'd be permeating rapture, wonder, or even harmony and peace. So much for pheromonic saturation. What's fertilising the soul and drawing the body into those legendary feats of days gone by? Self gratification and broken dreams, man. We're living in the era of die-hard realists. And the music's giving me a bloody head-ache, as well.

Where's the cadence to dream—where's your stair way to heaven? So try incorporating a new plan into mainstream thinking. You're up against a lot. All those values of liberated independents en masse, bear down on you. You're up against the metaphysical splendour of an ancient order down trodden into hard-boiled thinking patterns.

The present paradigm probably draws its roots from centuries ago. Perhaps it derives through St.Augustine's idea in the 5th century. His physical universe came into existence with time and not *in* time. I think Einstein's work gives this theory the paradigmatical representation we now use. His spacetime concept and discoveries seem to support that proposal. Well, that's fine. It's done a lot and we've come a long way. But I'm suggesting we've got eons and eons to go. And be that as it may, I'm not trying to pull down the barriers either. The fences aren't there for nothing. Look at applied science. Lasers and transistor verify quantum mechanics. Discoveries are happening, one way or another, through the spacetime and motion boundaries of the prevailing paradigm.

Science today however, involves uncertainty, gaps, finite thinking patterns, and no clear direction. Look I know Einstein's theories prove correct for the solar system. I know quantum mechanics realises amazing technologies. But these two major strands of thinking are incompatible. There's a lot of disparities between quantum mechanics and general relativity. For this reason alone they could limit greater scientific development. So I'm questioning the paradigmatical boundaries. And there they are—the speed of light, the beginning of time, the big bang and an expanding bubble perimeter. General relativity and quantum theory could align beyond all that. They could meet up across the road. With your bigger paradigm, science could address them together. We might pick up the dynamic attributes of two systems and the continuity of both within an even greater one.

Theoretically, micro and macro transformation processes involve symmetrical invariance patterns that create order. These patterns could be subject to a greater order of events. Taking the bubble paradigm full circle is a start. It's not enough though. You're still subject to the beginning of time boundaries. You don't get an entire nonlinear chain of events with it. My point is; there should be cyclical links through eternal transformation processes that neither general relativity nor quantum physics can explain under the present paradigm. Rather, they both seem to explain different corners and borders of a paradigm framing incomplete segments of reality.

. the paradigm limit of general relativity

Both general relativity and quantum theory seem to address different dimensions of the same paradigm of reality. One describes a macro level of spacetime and motion and the other, a micro. The macro revelations show a sweeping linear dimension of immense proportions. Within it, you get bits and pieces of nonlinear systems. Well, these little sparklers could still be subject to the linear lampshade of macro gravity. Now most people probably wouldn't suggest that. We're looking at highly sophisticated brilliant theories. They both reveal amazing insights into cosmic mechanics. Even so, I reckon both reflect the paradigmatical possibilities of partial reality. Your linear sweep doesn't cover a complete cosmic cycle. And it puts a blanket effect on the intricate systems within it. So what are these limitations? And how did this paradigm come about? Because general relativity precedes quantum theory I'm looking at that first.

It looks like Einstein stepped into the unknown with an ancient idea. Whether it was primarily through St.Augustine's influence, Newton's, or divine intervention, I think Einstein put the paradigm in place. It looks like he framed a causal nature to reality extending beyond the observation and experiment of the time.

Now I'm not an Einstein historian. (Hope you're getting used to these disclaimers by now. Cosmic Wonder; a theory of everything written by someone who knows bugger all about nothing.) But I've got an objective to meet. So I'm delving into his history to pick up on his paradigm. Because of Einstein's theory of relativity science operates within a deeper and more profound paradigm of spacetime and motion. Physics addresses more powerful energies, velocities and greater distances.

He took us way out there. Way out there though, still anchors firmly for us. He still began from the age old beginning perspective. He still starts from one point in time. He's blowing up our bubble, sure, but bubble it decisively is. It's a pretty powerful paradigmatical border line. And because he opened up a working model of the cosmos that explains buckets of reality, we think it covers all of reality.

Boundary number 1: we have a finite beginning. The spacetime concept underpinning general relativity seems to have one. It's creating a working model for science that disallows any reasoning of spacetime or motion before a specific event. The concept shows space beginning with time. Motion follows afterwards. Boundary number 2: motion neither precedes spacetime, nor appears as an integral component of it. We're getting a description of the pre-domination of gravity on the intergalactic scale. Can you see the linear sweep emerging?

The general relativity of gravity shows a limit to how fast everything under its guise can travel. Nothing goes beyond speed c: boundary number 3. Yet everything can develop into infinitely intense curvatures. (Escape hatchet number 1.) Anyway, with these boundaries you don't get your symmetrical barriers, or any other form of diversion. They're blocking us in accordingly. You're riding the linear plane of gravity within a cosmic phase of expansion. And you're not riding that far with it anyway because it goes pretty damn slow. Here we have a system of space and time ballooning out of nowhere. Boundary number 4: we have endless expansion. On one side of the scale it appears to be endless and open, while on the other there's 'the beginning'.

It's a linear framework. And it has an over-all circumference relevant to a bubble. The three possible formations of spacetime and motion through this paradigm all seem to reflect that. As partial formations of a sphere, they suggest that the *entire* universe is either expanding, contracting or flat. You're not getting the complete nonlinear convolutions of systematic changes according to it. Now these proven limitations form the perimeters of a working model of spacetime and motion, or the cosmos, through which science operates. It's a paradigm that enables us to determine plenty about the cosmos, however. So science can logically assume it's correct for the whole cosmos.

You wouldn't get the cyclical logic with the spacetime continuum; you know. You're only getting the linear reach. What we detect and know of gravity, looks like a gross linear part of a greater nonlinear framework. If so, it's not the complete pattern. It's only part of the picture. Theoretically we're only detecting the linear dimensions of the gravitational force formation, the orbital. And we could be extrapolating this one element of a more complex set-up onto the entire cosmos. Either that, or we're proposing chaos beyond this expansion.

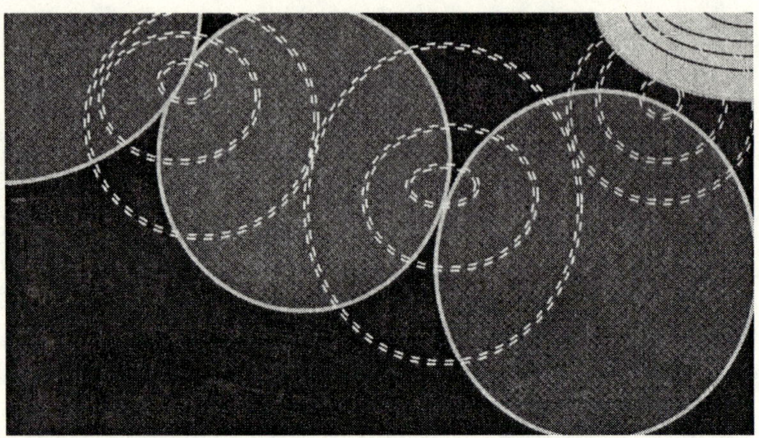

The pink area in that diagram represents the paradigm science uses today. Here we've got an expansion of spacetime and motion. The rest of the diagram is your nonlinear chain of events. Look at the light blue circles emanating from one of the green cycles into our paradigm. Is this the missing dark matter that gravitational theory *predicts*, but can't illuminate? The gravitational influence on galaxies shows the effect of mass that's not visible. This mass could transcend our observation below speed c. Well, those circles are the reverberations of an adjoining cycle. They herald the onset of our contraction phase. We can only look backwards or forwards so far through our paradigm, however. And those blue circles of mass would exist beyond, if besides, the expansion of our observable galaxies. They should exist according to a different big bang. But even if they're dancing directly to the beat of another drum, that drum's rhythm is in the same orchestral movement as your neighbouring cycle. Now there's a thought. Imagine the racket. The cyclical sounds of the galaxies; a cacophonic mind blowing symphony of melodic comings and goings. Or it could amount to the sweetest sounds any soul could reason. The sublime rhythms of spacetime and motion in symmetrical harmony could overwhelm the senses more than any fun ride of an IMAX screen. Who knows—I'm not writing songs, anyway. I'm just procrastinating off the point.

Theoretically gravity is subject to three other combinations of spacetime and motion. For a system to function and transform according to a pattern of symmetry, four combinations of spacetime and motion interact. They can't all be swamped by just one, all of the time. Where's the balance in that? Why should gravity override all else until kingdom come? On the macro level, the interaction of the gravitational orbital with the other three forces would show up as a colossal cyclical chain of events. Your big bang event, that paradigmatically kicked off the space and time of this gravitational expansion, is analogous to a nuclear spherical spinning formation. And you could have some quasar waves, and accreting spirals in between. Who's going to see it that way though? Not while we're subject to the speed c limitations of the gravitational paradigm, you won't. Ostensibly, spherical spin has minimal time and space with the motion of speed c squared. And the other two formations figure at speed c and above. So about 75% of motional factors of your cyclical scale transformation go beyond our boundaries.

In the previous diagram we've got several cycles happening. Where one area expands, another contracts. It's fits eternal infinity. You can see a pattern emerging that conceivably has no decisive beginning or ending to it. Here's our greater nonlinear framework. And it should clearly surpass any macro gravitational expansion. Going by the drawing board, that is. You see; by extending the paradigm we can reason a raison d'être for the expansion. Theoretically, it's a symmetrical differentiation. No longer is it unconditional. I don't see how it can be, if these forces are systematic. You need their four formations to account for cyclical nonlinearity. And all up, your bigger paradigm shouldn't invalidate the smaller one. Rather it should contextualise it.

Theoretically, gravity is a symmetrical composition of spacetime and motion. It's only symmetrical, and therefore systematic, in so far as it interacts with three other force formations. But our gravitational orbital formation has the biggest quota of spacetime and minimal motion. Evidently you can set up a cosmic paradigm with it. It easily shows up as the most relevant and decipherable one to us. Not that we're capturing the full force formation though. But its linear reach is probably the most tangible one in our hyperdome of reality. So we're probably judging the whole cosmos according to it. These boundaries cover everything we can see. And that's just it. It's also putting the blockers on inherent nonlinearity.

An over-arching paradigm of gravitational general relativity wouldn't just delimit the bigger picture. It's riding rough shod over your bifurcating possibilities. Your linear sweep is blanketing out key elements of any nonlinear systems within it. Theoretically, force formations and their symmetry replicate on different levels. That being the case, on the quantum level you could be getting entire nonlinear formations and processes of events imperceptible to us in macro reality. Physicists and theorists could be addressing the nuclear forces of micro systems through a linear plane of a macro system. You wouldn't pick up the superluminal possibilities under the bigger gravitational umbrella. The point is though; they could be comparing the detectable dynamics of one system to that of the other. Gravity calculates as billions of times weaker than the other ones. Well, is that according to its macro counterparts? Or is it according to a micro set-up of nuclear forces? I bet it's not billions of times weaker than your big bang. Yet the way we rationalise the gravitational force it looks as if it's weaker than the other forces by a 1 followed by 36 zeros.

The whole point of force formations is their symmetrical balance. We have a definitive motion limit for each one. This affects the space and time for each. In toto, they act according to a system of four. In toto, they amount to one force of spacetime and motion. But each combination, or force formation, sets a pattern of invariance that enacts transformations according to its system of four. The symmetrical application of the spacetime and motion of gravitational orbitals, for example, as we know it through general relativity, realises massive attraction. This macro version could be, for all practical purposes, indecipherable inside the atom. That's another system. It's got its own symmetrical set-up happening within. And your greater gravitational pull could be confusing your atomic reasoning. And vice versa. You've got all of those extra zeros to consider for starters. I'd save them for the bifurcating logic.

And if you think that's confusing, try the Uncertainty Principle. What a joke—there's something in my favour. Nonlinear rhetoric is a piece of crystal clarity by comparison. Inside an atom, we could have a micro replication of a gravitational orbital. Check out your electronic orbitals. Here it would be interacting with its micro mates according to its own little nonlinear system. Already I've gone into this. Your micro relativity of a force formation doesn't predominate the same way its macro counter-part does. Not that is, according to the bifurcating logic of systems. One branch carries lots of twigs rather than vice versa. And I'm not spelling out the woody logic of that one, so I'll skip the metaphors. Anyway on your micro level, gravity is diluting the nonlinear intensity of maximum motion in minimal spacetime, by providing an orbital quota of spacetime in minimal motion. It could be stretching out the motion mixture within a nonlinear situation so that nothing can travel beyond speed c through it. But here it's *functioning concurrently* with other force formations. It's not ruling the atomic roost. And this is what's fascinating. On the micro level, gravity supposedly functions within systems where spacetime and motion relates beyond speed c. Are you with me? We've got superluminal spaceship possibilities here.

A gravitational force formation on the micro level can, for example, simultaneously interact with other combinations of spacetime and motion. That doesn't appear to be the case of the cyclical level. Inside the atom, you've got your waves of electromagnetism forming with orbitals. And waves should also transpire by way of spirals and the spherical spin of atomic nuclear forces. So we could be getting the entire nonlinear pattern of events here. But who's going to see that through our paradigmatical limitations? We'd only get certain aspects of these processes, not the full Monty.

I think the paradigm of macro gravity only offers us the less transitory effects of micro systems. It should only affect ther atom per se, not its sub-atomic activity. Why don't we get position and momentum simultaneously for sub-atomic phenomena? We wouldn't pick up the active spiral involvement for starters through the greater gravity rationale inside the atom. Sure, you might get your spiral imprint on a bubble chamber shot. But that's freezing the frame of action. Spirals involve more intense motion within spacetime, than gravity. Theoretically electrons are symmetrical reactions to the intense nonlinearity of nuclear spherical activity. And spirals slot in between them. We'd only capture that below speed c. Your motion might freeze into a position at set speed.

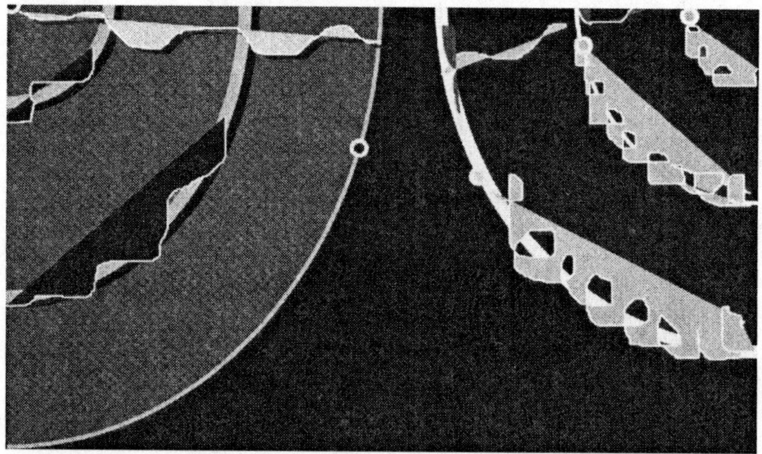

The diagram above shows electrons in orbitals around a proton. The left version is what we'd get through the general relativity paradigm. We either have the particle position or the wave movement. The right side shows the same situation using a different paradigm. This one recognises the nonlinear dimension. It describes the systematic set-up of four force formations, some of which exceed speed c. You'd get both position and movement of electrons together.

. paradigmatical limitations of quantum theory

Saturday morning, and no one can express quantum theory according to general relativity. Meanwhile, the traffic is droning. The houses and streets are happening. Sooty is lolling around in the shade, hanging out for a lizard. Is this obedience or has he lost the urge? Coming inside so soon? There's only 4 days to go my little love-dove. Then before you know it you'll be jumping and jiving with the butterflies once more—jumping and jiving with me around the barb-e-que, gazing out at the galaxies and watching the flames dancing. And we'll have quantum theory and general relativity aligned in a jiffy. Well, that's after we've further relaxed on the verandah, de-stressed and breathed in the mountain air some more. For now, we've got these peaceful grey walls through which to contemplate both theories. They're feasibly subject to the same paradigm. And if so, why are they so irreconcilable? It's not as if we're trying to work out your mid-life marriage, or something.

Already I'm suggesting the general relativity paradigm censors these nonlinear systems. It doesn't provide the depth of dimensions in which entire nonlinear systems function, particularly the depth of motion. Without perimeters recognising motion up to speed c squared, you wouldn't go full circle with your energy and matter transformations. Nor could you gauge the symmetrical replication of entire systems. Well, not according to this theory you wouldn't. Look I'm not going to get this done, if you have to inspect every squawking sea-gull out there. You're either in or you're out on your own, tethered. O.K. Sooty? And if you're staying up there, little nosy parker, don't you **DARE** climb the blinds.

Everything, in some form or manner, is theoretically subject to the symmetrical invariance of four forces. They take us up to speed c squared and back again. Our theories and laws of physics could be illuminating different parts of these processes on different levels. And neither part on either level is breaking down the barriers. On the one hand, quantum theory seems to explain elements of an entire system functioning. But you don't get complete continuity within it. On the other, at the macro level, we could be getting the continuity of only half a system. You get the linear span but not the nonlinear totality. So while each may be true to a degree, they wouldn't (a): correlate with each other, or (b): invalidate each other. Both could be incomplete. Neither seem to show the full symmetry of a transformation process through the interactions of four forces. And again, I reckon both are subject to the same paradigm.

So what can we find in common—besides the paradigmatical limitations? Going by the bifurcating logic, our micro and macro processes wouldn't precisely replicate each other anyway. I mean even if we were comparing the full macro cycle with the atom you'd have differences. Why have sub-systems, if they all function the same? Within nonlinear systems, gravitational spacetime and motion is only general in so far as the system allows. And get this. In an atomic system, micro gravity could have a specific, rather than a general relativity. Step outside your atomic shell and sure, you're more than likely not going anywhere beyond speed c. But inside it, you could be spinning around at unbelievable velocities. And we should still be able to correlate something about those dynamics with the greater gravitational reach. Although we're not getting the entire systems logic with our linear reach, it still expresses part of a nonlinear event. Take the big bang. The nonlinear intensity of big bang imbalance could analogise the instigation of an atomic charge system, for instance. And when we look at the on-set of radiation, we could be looking at a miniature version of an intergalactic critical density factor.

What then correlates to your quantum of action on the grandscale? Everything should in some sense. We've got stars, planets, galaxies, the moon and my foot, for starters. They're all discrete packages of stuff transpiring out of spacetime and motion. They all derive out of a systematic transformation of energy and matter a la four forces. The moon is a discrete quantity of stuff according to the stellar system, and my foot—some other one. Well, O.K. so I've lost the plot a bit on that one. But the point is, you can see packages of matter and energy in motion on various levels. The difference with our quantum one is that spacetime comes over as more dynamic. At this micro level, motion is more evident within spacetime. Theoretically we're at the bottom rung of the bifurcating ladder according to force formation systems. We're at the cutting edge of spacetime and motion.

Remember, with the atom you're theoretically getting the whole kit and caboodle of nonlinearity in a nut-shell. Atomic forces for example, calculate as quanta with integral spin. Inside it, we're looking at the integral nature of motion within spacetime, which can still form a discrete unit. That is; a quantum of spacetime *and motion*. And there's no system smaller. None that I can figure out anyway. We've advanced through the tree of general relativity to the leaf. Any further quantification is just dissecting the tiniest four force set-up. You're just tinkering inside the bottom line. Our quantum of sub-atomic spacetime and motion represents a minimal act of symmetry through a force.

Now theoretically, the relativity of motion, according to the spacetime of greater gravity, doesn't penetrate the atomic shell. Inside, your bottom line nonlinear system of four force formations is irreverently doing its own thing. Greater gravity—and its paradigm just over-rides our reasoning of them. So we can't grasp anything beyond the speed c dimension, *inside or outside* the atom. We get the wave motion at speed c, or a particle position in an orbital below speed c. But note, we're still getting our orbital in the atom. That's because it's the same spacetime and motion combination on any level. We're still within the boundaries of our paradigm with it.

The thing is, once we've calculated our sub-atomic ball of spacetime and motion—according to the bottom line mass and energy transition, we can't move with it. Too bad, we have to forgo the motion for its rational position. There's our prevailing paradigm stepping in again. It stops us seeing the position and motion simultaneously. We're viewing all activity here through the lenses of greater gravity. And your linear sweep doesn't recognise the mass-motion fusion of intense superluminal velocities. So you can forget about your spiral and spherical spin interactions as well. They too are out of vision.

The sub-atomic photon, for instance, has zero rest mass. That is; it *calculates* as having zero rest mass. This could be our theoretical mass-motion combination of nonlinearity. As per the linear dimension though, it would register according to motion *or* mass. The photon furthermore, should have spiral affiliations. But if we ever see one sub-atomically, it seems to be the remnants of something very transitory. Theoretically, spirals signify an even smaller time span for a quantum of space and motion than the photon of wave. It's got a greater motion ratio. That means the spiral quantum of energy has more motion because it has less time and space. No wonder we only get the trails of transience.

Anyway, what transpires between the strong nuclear spherical spin and the wave formation, should involve some amount of weak force interaction, or spiral activity. Inside the atom we've got four force formations firing off each other. Indeed, there could be a lot more to radioactivity and electromagnetism than we realise. At least by way of the greater linear dimension. Where's the continuity of such processes and interactions according to our paradigm? Our growing quantum zoo isn't joining the dots. Despite their input we still don't have a clear picture of sub-atomic activity. These strange little numbers only reflect an increasing quantification of data that doesn't quantise according to known values. In short; our greater paradigm, that of general relativity, is deflecting the brilliance of inner nonlinear systems. It's rendering quantum limitations. We're looking down upon the night lights of New York through a smoggy shield.

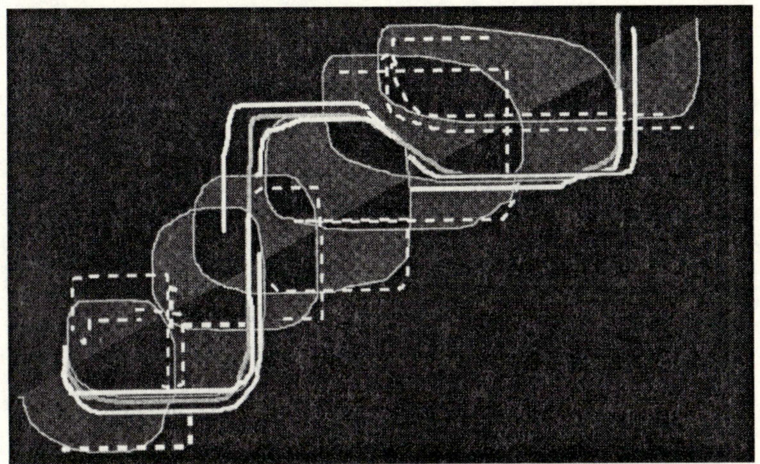

The above diagram shows spiral activity. Superimposed upon it is a wave formation at speed c. Now you don't get all the dotted spin factors with the wave. The wave only depicts what is definable according to general relativity. On that basis there are the yellow lines of what could be an up-spin of say, a charmed particle. Maybe the green at the top translates into the down-spin of another quantum of activity. Whatever; the point of this repeat diagram is to illustrate how spiral activity could transcend, and translate through our present forms of measurements.

. can we link them within a broader framework?

Apparently there's a crowd of people trying to find a unified theory of physics. I guess it's taking on the spectre of a holy grail. Whether it's through orthodox or unorthodox methods the objective remains the same. Whether it comes from within established scientific circles, or not, the challenge beckons. Well, it's not beckoning me. I'm not especially trying to produce a 'grand unified theory' of anything. Scoff, if you must. Phooey to fame and fortune you think? That's not the challenge either. Fame and fortune can come out of a soap opera. The challenge here is to work out ways to get beyond speed c. The challenge is travel the galaxies. Enigmatically however, the discordance of these two theories seems partly why we're not there. Aligning them seems par for the course. If so, a grand unified theory might eventuate along the way. You see; the superluminal dimension is supposedly within the same grander context that ostensibly unites these theories.

Already the speed c barrier, as a tool of analysis, seems to be producing particular conclusions regarding the inadequacies of general relativity and quantum theory. Indeed, this search involves the many avenues of mystery within cosmology. And just when you're about to walk down one of them, a blasted racket kicks off shattering the elusive horizon. Those earthly realities draw you back into the bubble where you belong. And what promised to be so revelational now gels into a figment of fairy-land ostrich mentality. What do you get instead? Another war, that's what—with your home country in the thick of it this time. We've apparently traded in the Kosovo crisis for an Indonesian one. So now I'm wondering what the hell is the media without a war to report. When is our war intrigue going to transpire into disgust? Perhaps we can push it off the headlines with something more powerful— with something good for a change. Now there's a challenge. The possibility of superluminal travel might illuminate a greater future than frigging fighting does. But who generally cares? What's exciting about inter-galactic travel without all the high powered star wars' violence along the way? It sure packs them into the cinemas.

All right then, I know there's the realities of independence for East Timor. There's no denying the atrocities, angst and terror going on. But since when is statehood an easy transition? Nine times out of ten it involves a bloody battle. But that's also the gory entertainment factor, I reckon. Keeps you tuned in and 'aware'. Why haven't they figured it out from all the other bloody battles for independence? Why didn't the international input happen before this latest one kicked off? And I bet by the time you're reading this another one's going on somewhere. Well, the media's there all the way—pushing and shoving, vying and contriving until whammo, let those cameras roll, we've got another full blown performance happening. You've got to blame someone haven't you?

Well, bugger that. I'm heading down one of these mysterious avenues. I'm staying focused cosmologically. My little honey pie is curling up—all warm, cosy and safe and sound after another big night out with the wallabies. What a blessed relief to be overlooking the valley again, surrounded by mountain ranges of trees instead of traffic and trains. I'm just soaking in their misty variation of green and grey into the horizon. So let's do it. What this speed c journey is showing are even greater avenues for analysis then first meets the eye. To link the two major theories within cosmology is possibly just a drop in the analytic ocean. And we're still only setting up the paradigmatical scene. Then there's all that bifurcating logic to explain. And how do we rationalise cyclical links on the broader scale? How do we know these are the systems through which the cosmos functions, anyway? We don't. I've got to delve into them some more. I don't see how these theories can link in with each other unless we can also fathom what happens before the big bang and the point of our expanding universe. And that's until we clarify our greater paradigm.

So far we have 3 discernible paradigmatical limitations; the speed of light, the beginning of space with time, and the inexplicable expanding universe. The main thing that seems to apply to *both* quantum theory and general relativity at present, is this speed barrier. On the small scale we have gaps and uncertainty with speed c, whilst on the large scale we still have question marks. Perhaps the speed c barrier contributes to those other limitations. What was there before the beginning of expansion? You can't tell below speed c. How can the gravitational mass we detect cause such expansion? What comes next? Does the bubble burst? Who knows when you can't see that far in time. One thing is clear though. On neither scale do our theories and laws of physics cover an entire systematic process. There's no simple guidelines applicable to micro as well as macro phenomena.

As science stands today, I think you've got Buckley's chance of constructing a paradigm or working model relevant to both areas. Quantum theory and general relativity probably influence each other. Their individual truths, powerful as they are, could also have push and shove. They could be directing scientific thinking into opposite directions with tremendous heave ho. Our resulting technologies are probably going up different avenues within two contexts. Imagine what we'd get if they align; clarity of vision for one, a better sense of direction and our place in the grand-scale scheme of things for another. Why, for instance, do we need to find an enormous amount of dark matter to explain the gravitational behaviour of galaxies and their halos? Why can't we understand such behaviour through the comparative strength of four forces? Do we completely understand the comparative strength of the four forces? How do we know the relative motion of the galaxies is not subject to large scale waves and spirals of weak nuclear force formations? Their motion might be subject to the macro counterparts of the sub-atomic nuclear forces, in particular the symmetry of them. Indeed, the non-luminous matter we're searching for, to explain galactic movements, could be the stuff of superluminal velocities. Stuff that is pertinent to, and analogous of, the strong and weak nuclear forces of the sub-atomic scale.

The thing is; to prove the expanding of big-bang theories we require invisible dark matter. Apparently the gravitational influence of visible stuff is insufficient to explain the stability of fast-moving stars and galaxies. So it looks as if there is an unknown effect—a great one, on the movements of the stars and galaxies. Furthermore, it's one that we can't account for through general relativity, or for that matter through quantum theory. Indeed, according to our theories this hypothetical dark stuff could amount to as much as 95% of the universe. That being the case, we could be understanding only 5% of a system, or what is evident below speed c with our laws of physics. So does this mean the linear aspect of a nonlinear system amounts to only 5%? Sorry. Some inadvertent discordant maths is apparently creeping in here.

Well, this invisible stuff scientists invoke to support our present theories, could involve the superluminal activity of 3 forces plus gravity. More specifically, it could involve an entire neighbouring cyclical set-up of them. How would we see it that way though, if science pursues different avenues within the micro and the macro? That situation only seems to render forces practically incomparable. You don't get the systematic set-up of four of them on either level. So you wouldn't arrive at the bifurcating potential of a cyclical scale of them. Again, I think it's something akin to comparing the linear with the nonlinear. On the one hand, we could be reasoning the linear element of an entire nonlinear system. On the other, we could be comparing it to segments of sub-systems within it. We might not be gauging any one force's strength systematically, according to its entire nonlinear situation.

Clearly there should be one set of values relevant to all four forces on both the macro and the micro levels. And now we're possibly getting to the heart of things. *Spacetime* seems inadequate to form the paradigm for a uniform quantification of the cosmos. I don't think it's the basic *quantizing* tool of the universe. Sure, it might suit the general relativity of a cyclical expansion stage. But that's my point. That's all it is. The spacetime curvature of our paradigm is just the macro gravitational effect. I'm suggesting; the four forces, and all their consequent transformation processes within the cosmos, can extend beyond the boundaries and values of this spacetime curvature. That's on any scale. Look, there's no denying the realities of the spacetime curvature. It's stabilising a great whack of cosmic bifurcation. It's just that; conceivably—theoretically, there's cosmos beyond it.

In short, we need a basic quantizing tool for a greater cosmos than we presently envisage. I reckon it's rotational motion. And rotational motion, I think, is a matter of spacetime *and* motion. We're opening up the spacetime paradigm with it. With motion, spacetime can stretch from here to eternity. It covers every possible cosmic curvature. It goes with the flow. And it forms the basis of a nonlinear system, thereby leading to a systematic turn of events. Theoretically, you can go with this into every outpost, dynamo, and eagle's look-out. From inside an atom to the back of a quasar, stuff figures in some form of spacetime and motion. It figures in some manner of a force formation system. Even with our limited observational abilities, you can see it. Rotational motion appears typical of most structures in the universe. You get it from neutrinos to galaxies.

And if it's not rotational motion, what is it? Sure, I might be barking up the wrong tree, but if it's not this tree I reckon one exists that grows everywhere. Clearly, having one set of values applicable to everything in the cosmos should simplify cosmology. We're not just talking superluminal potential with this. We can light up every nook and mysterious cranny with it. This is about discerning order out of chaos. We can systematise everything uniformly. With a proven set of values, we can confidently extrapolate into eternity. O.K. then, so we haven't proved it. What does it take? The signs are all there. The cosmos is out there in front of us— stellar systems happening, atoms going on. Just because we've deciphered two major strands of physical laws that are unaligned doesn't mean everything is unaligned. Who are we to suggest the universe is unaligned, anyway? We probably know bugger all of what there is to know. Plainly, there should be values applicable through out any level of transformation processes within the cosmos. How an atom functions—that should have some relevance to how a stellar system functions. And that should likewise relate to how a galaxy happens. And what's so far-fetched about that?

Logically, all these systems should concern four forces and theoretically, their own nonlinear dimensions. Already I'm suggesting four force formations make up a nonlinear system. Now take it as a paradigmatical pillar of strength. Then take it further. Open up the paradigm. One nonlinear system might apply at an intergalactic level. Then they could bifurcate progressively down to the sub-atomic level. Or don't even contemplate it. Stay in your bubble, and curl up with the safe and sound laws of physics to date. Lord it up in today's wisdom over yesterday's, and stuff tomorrow. Tune in to and turn up the latest flash-point on the planet. Revere the known and ignore the tangled confusion beyond. And don't wonder why. Well, I won't be, if I don't go and help finish digging the trench for the power cables. Some luxury it is to sit here and contemplate the ins and outs of what might and might not be cosmologically the universal case, while the car battery is on its last legs. And that's not to mention Ken's back—and temper. Nothing's ever easy is it—not writing a book, not wiring up a property, and try raising a garden in gale force blasts. Equinoctials or minstrels, whatever they are, here they'll stay for the whole jolly month no doubt. Go on blow, see if I care, you're not talking me into a wind-mill.

The limitations and inconsistencies we've been confronting so far, could simply be symmetrical tangents. As different combinations of spacetime and motion, we've got four forces that each have a formation. We've got four different curvatures of spacetime and motion—our basic value. As such, they should act according to each other for systematic cosmic organisation. What's the purpose of differentiating spacetime and motion anyway, you'd think. They should balance each other and together make up a 'whole'. That means gravity should act as a symmetrical counterpart with the nuclear forces, etc. So we just need to discern their system. And with our common set of values we can do this. Together, they can light up the systematic set-ups of cosmic processes. You see; the combined symmetry of force formations amounts to an entire non-linear transformation process.

Moreover, their individual values still apply to—and don't change with different scales. Here is a common set of values between the micro and the macro scales. A force whether magnified or miniaturised, would always take the same formation. It's always subject to a specific balance of spacetime and motion. And it's the same one on any level. Systematically, however, they have a progressive organising effect. There's your atom firing away correlative to your stellar system, etc. But I'm proposing the atom's firing on all fours at once while your stellar system is a more gradual affair. I mean you probably wouldn't decipher four force formations happening simultaneously with your stellar set-up, whereas you might within your atomic one. They're supposedly interacting according to their proliferation. Eventually however, these transformation processes conceivably amount to a simple skeletal large scale cyclical pattern that carries forth into eternity and infinity. Now if that doesn't form the mother of all paradigm's, what does?

In case I've missed something here, I'd better summarise the main points thus far. Our four force formations interact according to the same formation, and combinations of spacetime and motion—on both the sub-atomic and the macro levels. Their types of interaction however, whether they function simultaneously or otherwise that is, depends on their systematic size and its proliferation. But the main point is; spacetime and motion differentiates into four combinations or forces. And despite the size, their values remain constant. It's a metaphysical grid. That's the theory. These combinations, or forces, act and interact according to their combined symmetry or formations. It sets off a pattern. We have more transformation processes, and so new ratios of energy and matter, etc. And it all amounts to the bifurcation of systems—more or less.

Take the Mandelbrot patterns of chaos. It looks like the replication of symmetrical patterns to me. Indeed, where's the chaos? Mandelbrot seems to have captured an order to nature here. All we need to work out is the finite framework of replication. I'm just proposing it's a system of four formations based on their symmetrical balance. On the greatest level they all come back to a pattern of just four. And this large scale cyclical pattern forms internal systems of transformation processes. It's a two-way process. From their symmetry further transformations of energy and matter occur. But they all interlink, and at some stage, the larger scale symmetry puts the blockers on further internal bifurcation.

It's a simple set-up when you think about it. Nothing chimeric to it. Because we have four distinguishable forces, it's possible there is no *one* curvature of spacetime, but four formations of spacetime *and motion*. These four forces conceivably interrelate, and at some stage unify. They should all have powers of spacetime curvature. Why should the entire cosmos always be subject to the spacetime curvature of just one of them? I'm suggesting we're not always subject to the general relativity of gravity's paradigm. Space, per se, might not begin with time and curve into motion consequently. Perhaps it only looks that way according to our over-arching cyclical level of gravitational predominance. That is, according to the prevailing paradigm.

So where do the primordial curvatures come from, you may ask, if not from the all time big bang? Does spacetime and motion basically differentiate out of the blue? Hell, no. Theoretically, these forces should arise through the energy and matter ratios of a functioning *system*. And if you think I'm talking in circles—good. That's what it's all about. This is an ongoing never-ending process. Certain levels of energy and matter effect a forceful response. I'm just saying it happens systematically. And it's a two-way thing. Your differentiation of spacetime and motion into four forces feeds off energy and matter ratios, and vice versa. And look, I'm not going into and over this much more, by the way. You think you're going around in circles—I'm sitting sideways in the chair with it.

Everyone knows energy and matter are always transforming. I'm just saying that's because they're subject to spacetime *and motion*. Further, that symmetry is the basis of change. Every transformation is supposedly subject to four forces of symmetrical association. And every action has a symmetrical basis, no matter how indirect. To understand a cosmic order here, I'm recognising the *systematic* turn of events. Your common cosmological value leads to it. When you decide upon one, you're up and running with it. Well, you are when it concerns necessary motion. We're talking the method behind the madness now. So it's not simply that spacetime and motion are the ingredients of energy and matter. Rather it's their symmetrical happenings. Theoretically therefore, symmetry effects a reciprocal response between four force formations of spacetime and motion and energy and matter ratios. This business of balancing one with the other should lead to systematic changes within the cosmos. In other words, given x amount of energy and matter according to change, an imbalance can occur. Different forces kick in to maintain systematic equilibrium. And so it goes. It comes and goes.

On the cyclical scale, for instance, the expansion phase would eventually contract and converge into another cycle. And at this massive level, you're looking at the predominant activity of one force formation at a time. In your sub-systems, remember, forces theoretically operate more in tandem. But at your cyclical level, you can't pare down your force formation interaction any more. They become subject to the adjoining cycle for the formation to figure. And your grand-scale symmetry of them can rein in all the internal bifurcation of sub-systems—the stellars, the atoms. They all contract under gravity's predominating combination of spacetime and motion below speed c. Then they might all surge through a quasi stellar set-up at speed c. Then they might accrete through a predominating spiralised force of spacetime and motion. Whatever. They're all subject to the next grand-scale force formation that figures at this level.

Look at it this way; this cyclical scale could be the symmetrical benchmark of spacetime and motion differentiation. The all encompassing one-on-one general relativity of a force formation could be a cosmic primordial event. At least in so far as the internal manifestation of systems occurs. These theoretically happen according to, and within, the borderlines of each individual cyclical formation. That is, within the general relativity confines of your macro force formation. We're looking at the ins and outs of a fundamental cosmic slab of spacetime curvature. Your cycle is setting the stage for finite happenings within eternity and infinity. It's framing the Mandelbrot pattern of chaos. And what are you sleeping for Sooty? Look at those swallows out there trying to nest in the verandah. Listen to those bloody frogs in the fish pond—what a racket. How can you sleep through it? How can I think through it? Well, at least swipe the blowie for me. Don't then. Just don't expect me to worry about any red-bellied brown snake waiting for you, next time you go down the wombat hole. Cheeky, naughty.

Our greater paradigm, then, allows us to systematise general relativity. This is according to cycles. Here we have the spacetime and motion transcending a big-bang starting position. We've now got the space to extrapolate into the greater nonlinear dimension. Even so, such space is placing us firmly in the realm of theory. Our links with reality, whether drawn through a giant-sized paradigm or a monocle, could be insufficient to convince a physicist of greater gravity's future prospects. You've probably got more chance elucidating froggie wonderland to a tadpole. We're chewing up cosmos beyond present observation and experiment. It's that simple. Using present methods we stop at the black hole event horizon. Without superluminal technologies, I don't see how we could prove any of this. Without the nonlinear, numerical brilliance of a mathematical giant, who is listening? Would a picture help? Below is one massive gravitational orbital. It's bellowing out of a spherical spin, etc. We could be looking at the translation of a big-bang spiralling and stabilising through waves into a gravitational orbital. Eventually, the orbital should progress full circle into another cyclical link.

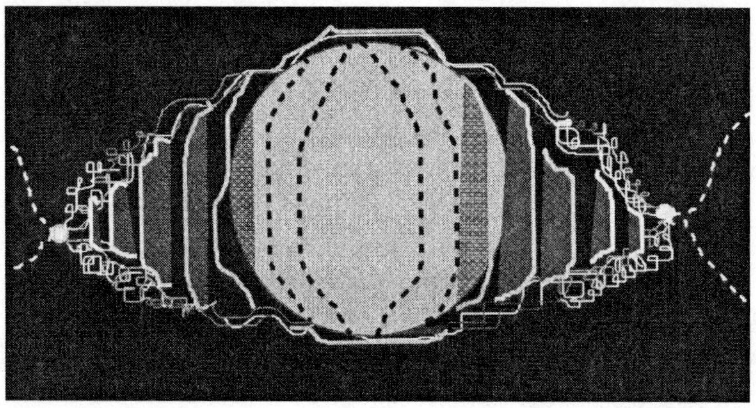

When you arrive at the boundaries of gravity's general relativity you're stretching the observational point. All I'm doing therefore, is drawing a gravitational expansion into the full circle of an orbital. I'm showing you the dark side of the moon. On that basis, macro gravitational stability is decisive in both the expansion and contraction. And to co-ordinate the formations on this level, they each have a predominating affect over the same chunk of cosmos. Pursuant upon each other they lord it over the same cosmic slab. Your dimensions of energy and matter ratios are symmetrically subject to the acting formation. Bearing in mind though, that the neighbouring cycle affects your formation as well. In this way, we have an inter-linking process of expansion and contraction that could amount to an entire nonlinear system. Our extra space afforded by a greater paradigm—one transcending the present laws of physics, covers the push and shove of your neighbouring cycle. This is the space that is instrumental in our contraction phase and thus the formation of a full blown orbital formation. So with some extrapolation, you can cover the mind-boggling nether-regions of the entire cosmos with it.

One such inter-galactic expansion-contraction cycle could encompass the size of 'The Local Group' and take 15-20 billion years. Who knows? I'm just hazarding a guess here. I don't want to. There shouldn't be any guess work. But the questions remain. What's the local spacetime necessary to extrapolate into one over-arching nonlinear framework of cosmos? How much do we need? How much do we need to symmetrically and systematically co-ordinate with the quantum theory of the sub-atomic arena. Do the 'Local Group' dimensions fit the bill? Could we gauge consecutively smaller four force formation systems inwards until we arrive at the sub-atomic level? Or we can start from the micro and work out. The fact remains however, that at some stage we're beyond the general relativity paradigm, and can only extrapolate from reality according to theory.

Even so, you should be able to pick up four force formations on the atomic, stellar and galactic scales. Well I can find some of them there. I reckon you can decipher a nonlinear pattern of events in all of these systems. Clearly, these formations exist on an inter-planetary as well as inter-galactic level. We see, for example, plenty of planetary orbitals and millions' of stars spherically spinning. If we can discern a wave and spiral occurrence, there's your stellar four force formation system. Maybe a spiralised disc predicates your spherical spinning sun. I don't know. But en masse, they should amount to the greater galactic scale with less copious formation interaction. 2n b Here we're looking at the pared down events of spiral galaxies, quasar radioactivity and all encompassing orbital rotation, etc. We're addressing the primordial skeletal system of bare essentials. Whereas theoretically, at the sub-atomic arena these four formations are almost immeasurably manifest. Indeed, when we reach the sub-atomic level, the inter-galactic gravitational force should be practically negligible, and to date, improperly comparable to the atomic dynamics of a nuclear one.

What I'm suggesting is there should be a progressive and discernible symmetry between the cyclical level of macro gravity through to the atomic systems. Take it as a symmetrical trickle-down effect. We can have varying nonlinear systems seemingly independent of each other. Yet they would still be acting according to a common pattern of symmetry. They'd still be subject to common values. Macro gravity has the same values as a micro orbital formation. The macro version is still subject to the symmetry of three other force formations. Yet the *systematic* interaction of gravity on the cyclical level wouldn't necessarily replicate that of the sub-atomic. How could it? There's no myriad interplay of force formations. This is an all encompassing force formation. Macro gravity is acting through a gradual orbital formation on a general level—whereas your micro one has bifurcated beyond comparison into billions.

Each massive formation though, theoretically bears down upon it's internal replications. Each macro curvature of spacetime and motion should play a symmetrical role not only effective within its own four force set-up, but relevant to the sub-systems within. I think therein lies our links for comparison. That is; your greater formation could affect your *sub-system* of four formations, more so than your sub version of the same formation. So your macro orbital, for example, wouldn't simply manifest a blanket effect of below speed c motion and spacetime curvature. It would tap in to the symmetry of the sub-systems within its massive curvature. As someone once said

"... the spiral arms are processes transitory by the spatio/temporal standards of the Milky Way—they plough the fields to spawn new stars" n b 3

. endnotes

1 nb How the spacetime concept relates to biblical wisdom is looking intrigueing. The bible seems to outline moral principles from the same starting position, Genesis and the big-bang, for instance. The spacetime concept won't alow us to rationalise anything beyond speed c. And if we eat from the fruit of knowledge we're susceptible to our own misfortune. And they were right. It's come to pass, to some extent. Nuclear power leads to Hiroshema with us. Imagine what we'd do with superluminal power. Maybe we still need to evolve some more in the bubble, before trying to reason stuff transcending beginnings. We don't seem to have evolved enough to handle anything beyond its framework. On the other hand, if we don't escape the bubble we could become swamped in our own realites and implode.

2 nb "waves, generated by the harmonics in the gravitational interaction of the myriad stars, move across the dish in a spiral pattern" ref.lost

3 nb another lost reference

ch.8

cyclical links

. preamble

Now we're up to the last part on greater cosmology. That's in so far as talking total cosmos. I'm wrapping up the hypotheticals on infinity and eternity. Ideally chapter 8 then, is going to be the culmination of a theoretical symphony. Realistically, you're probably getting the last straw of presumption followed by something stronger. I'm clearing the way for superlumunal brainstorming. And the weather's definitely lending itself to it. Or is it just a ploy for stock piling those big bottles of Boags for a sunnier afternoon?

Theoretically, infinity and eternity are the most powerful concepts in the cosmos. And that's God for you—omnipotent and only up for belief by us. But clearly just believing in infinity and eternity, be it divine or otherwise, won't get us off the ground. What's becoming increasingly probable however, is that it's a start. This premise presents avenues towards superluminal possibilities. It's a mighty spring board. The trick is to bounce off reality with it. I want to come up with some superluminal theoreticals through established knowledge. And equipped with the above belief you can dauntlessly draw on our knowledge to date and challenge it. You can explore pre-big bang events, for example. We can get into cyclical 'end-points'. We can contemplate transitions in reality rather than end-of-time and end-of-motion scenarios.

Your cyclical junction could also be the final bridge across the river of doubt. No matter how cynical and disbelieving you are, there's one thing that any die-hard realist can't deny. Our laws of physics don't explain everything. Indeed, I'm theorising along the crest of a tsunami with their short falls. Thus far, we've powered into the cyclical dimension. Here is the outlined scale for linking sub-atomic uncertainty with general relativity's limits. The thing is we have to go further with it. To prove its points we need to step into observational reality at the same time. And our cyclical links, if we can figure them out according to both theory and observation, could take us up the mental process ladder from belief into reasoning and understanding. Reasoning and understanding eternity and infinity, that is. Any superluminal possibilities might then transpire into superluminal probabilities.

In clarifying a new dimension for our laws of physics to progress through, cyclical links also outline the systems approach to the cosmos. Already I've tried to correlate a systems approach all the way down the bifurcating line. But I guess you could still achieve that—to some extent, within a big bang scenario. It's just that cyclical links carry forward the symmetrical extremities of spacetime and motion, and energy and matter. You're getting the whole cosmic picture with them. If we can carry forward the systems approach through them, there's no room left for random activity. No chaos.

How can one force singularly define all cosmic processes, anyway? Forces should act systematically for this purpose. Their system should form a symmetrical pattern of events. Theoretically, it all stems from the infinity through which they act. I reckon it's only the infinite and eternal nature of spacetime and motion that can produce this effect. Indeed, the nature of infinity seems to invalidate the notion of chaos. It presents the scope to pattern events into ongoing processes of transformation. It presumes order. I don't see how you'd get to that stage relying solely on our laws of gravity. They only take you from a big bang starting point—nothing's going on before it. And then you have an endless inexplicable expansion coming out of it below speed c.

You know my position. On the grand scale we can decipher matter curving space into orbital pathways. But is the orbital formation, and its hyperbolic derivatives, the only mode for cosmic processes? We have three other forces. These could also curve spacetime—and motion, to produce different pathways. And these pathways could all interact to form ongoing cyclical processes. At the same time, they could be manifesting intricate nonlinear subsystems within their colossal framework. And that's the theory. Proving it is something else. Still, who's going to *disprove* it. With gravity as a guide you're only getting as far as 10 billion light years or so. I'm dancing around the gipsy bonfires aeons away from that ephemeral foot print in time. The age of the universe, according to theory here, is well beyond that. It's eternal.

Look patterns develop and imprint knowledge. Habits set in. Tell me about it. Cycles display the effects of hereditary selection. You can translate all that to the cosmic unobservable. Mechanisms imprinting cosmic information might transcend our laws of physics. What egoists are we to presume otherwise? Know-all earthlings. There they are—still annoying the bejesus out of each other and their planet. Can't even figure themselves out. If we accept our universal expansion as the only way to be, extrapolation is probably a waste of time. But maybe not everyone accepts that. What about those suggestions of critical density factors in processes? They're fathoming patterns for protostars now. Meanwhile, the mysteries of distant galaxies remain in front of us to examine. So perhaps the basis for theoretical input and extrapolation does, after all, exist.

What I'm getting at is the dimensions of cyclical evolution. I reckon they depend on a greater scheme of things. Perhaps only eternity and infinity can account for them. Going by common knowledge; one cycle could take 50 billion light years. That's taking into account; the time of known gravitational expansion, allowing for it to continue somewhat, and then a contraction phase. Then imagine *systems* of them. The dimensions would be so vast you'd have to be presuming eternity and infinity. Then imagine coming up with an equation for them. How would you be, figuring out an eternal and infinite calculus that sums up the entire universe. Something neat and nifty. And here's my effort below. And so what if it's pre-school art work—you could shove that on a postage stamp.

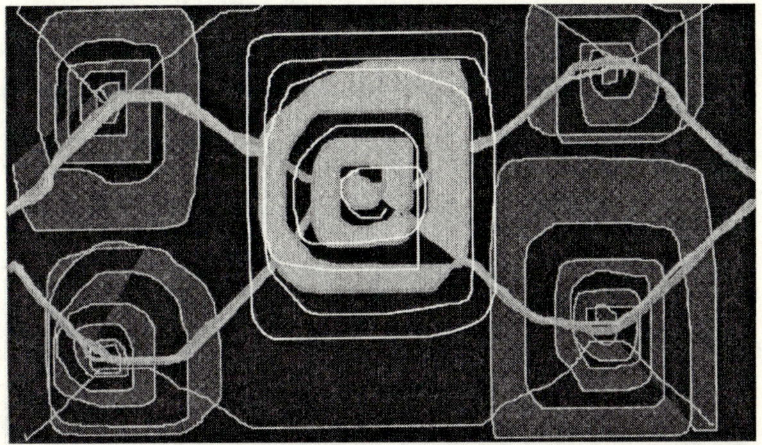

Here's our theoretical cycles forming. Emboldened in the middle is our one. It's forming within a system. It's contraction and expansion stages would be subject to adjoining processes. One cycle might expand so far, then start to contract according to its neighbours' position, as well its singular internal factors. The cycle as a system however, should resemble a process of symmetry.

Plus you've got teleological possibilities here. Contraction and expansion phases could leave an impression—a decisive impression. The pathways of one cycle of events might imprint and influence future processes. You know habits set in. The symmetry might reverberate through the annals of eternal time as well as the dimensions of infinite space. It's all gearing up to manifest the cyclical motion of the cosmos. The more you do it, the more you do it. Take bifurcation. Through the 2nd law of thermodynamics, bifurcation apparently appears as a spontaneous creation of order. Yet each time the forking, or division into branches occurs, so does a memory of a path come into play. Now if this process is a forward progression it could connote a cyclical sequence. That being the case, bifurcation could be a mechanism of symmetry.

The blue spiral in the middle of the previous diagram represents a memory pathway. These could be tracks laid down, inevitably, within the fabric of spacetime and motion. The blue pattern could be influencing our bright cycle. I'm even wondering that cosmic microwave background radiation has something to do with it. Can a quantum flux event rebound through the evanescent cosmic signals from aeons ago? I mean these things, nebulous or not, seem to permeate the universe.

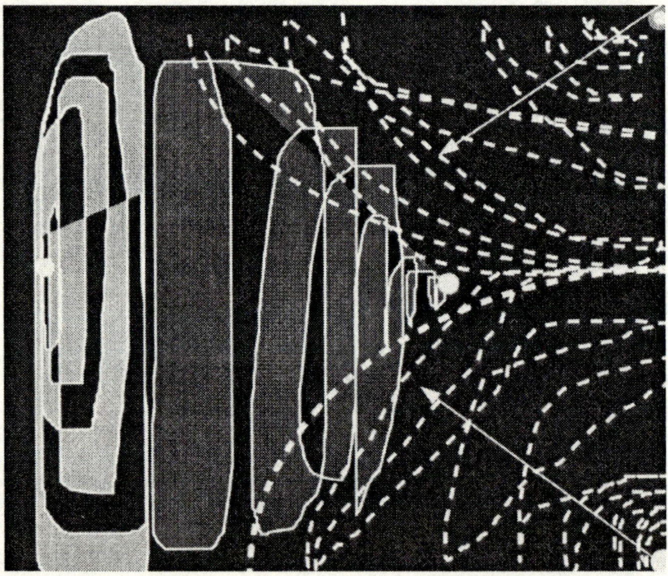

This diagram is another representation of the systematic process of cycles. Our main cycle concerns the dark blue expansion and the light blue contraction areas. The dotted formations are causal phases of older, distant, adjoining processes. I'm trying to show the contingency of a symmetrical relationship, where voids still have meaning. With symmetry and force formations we always have different combinations of spacetime and motion. Your interstellar void, for example, could be a buffer balancing two greater formations. Or it could be some internal bifurcating effect of the sub-systems. Either way, in some manner it's part of a formation. Any vast open space would be subject to processes going on further up the cosmic road as well as the immediate set-up. We're systematically mapping out the entire cosmos this way. So any cosmic area of seemingly 'empty space' is still contributing to the balancing act of spacetime and motion. But that's the other side of the coin. The issue here is the cyclical 'end-points'.

Thus far I've been pretty general about them. To decide on these junctions, I think we should arrive at one more specifically than the previous paradigmatical investigation allowed. I want to spell out the macro events of four force formations acting as a system. Do we see them out there? I'm going into black holes, quasars and accretion discs in this chapter. There's also something on cosmic rays and dark matter. How do these things fit in with the theory? And in what manner can they approximate sub-atomic systems?

Theoretically at the atomic, stellar and galactic levels, the four forces interact systematically. They proliferate progressively inwards from the greater level. So on the cyclical scale they're a skeletal framework. Here you have the barest essentials re force formations. But this framework involves cyclical links. And these should still correlate in some decipherable manner with our sub-systems' connections. We should be able to decide upon a systematic symmetry all the way down the line, by way of the cyclical set-up. That's our proposed position for comparing the macro with the micro. That is; I'm comparing one *system* of four forces with a sub-system, rather than one macro force with a miniature counterpart. And to arrive at this comparison you have to determine the systematic macro set-up of them. You have to reason the links in the cyclical chain.

Finally, cyclical connections I hope, will show that symmetry does not produce mirror images of time, space and motion. Invalidating time reversal scenarios is big thing with this theory. The superluminal presumption is that we cross over the speed c barrier rather than rebound. I don't want to bounce off the barriers of the laws of physics. I'm trying to transcend them. But our motion barriers seem to carry through to every possible level. If we don't decide on a cyclical junction, we could be back in the bubble nature of existence. If we can't cross over the gravitational expansion into another general spacetime curvature, one that involves another force—one that goes beyond gravity's motion limits, I don't see how we can reason superluminal motion. You'd probably end up back-tracking in spacetime through motion instead. Within infinity and eternity there's no need for that. Where's the cosmic full stops? There's no need to reverse in theory to maintain the validity of our laws of physics.

With this theory, we don't go around in the same circles. Nor do we expand endlessly from a set beginning. At some stage a new formation emerges. Just like people. And the formation remains subject to symmetrical processes. Further, a forward progression makes each pattern of events, to some degree, unique. It would be the nature of infinity and eternity, expressed through symmetry, that allows this.

Cycles, and the symmetries they presume, provide a measuring device for an 'incomprehensibly' vast universe. We can extrapolate forward and backwards with them. We can fit theory with reality. And that only seems possible through the greatest dimension going. Someone's grand plan you think? Or are cycles and symmetry a natural consequence of an infinite and eternal cosmos? Perhaps you can't have one without the other. Either way, this measuring device could take us beyond our present limitations. It can relegate the notion of chaos to the dark ages. It offers the scope for dreams and development and greater evolution—the unscientific forage that probably precedes any greater realities. Besides, it could also be a scientific fact. Cycles and symmetry—who knows? I might as well laud the theory. I might as well trumpet it out anway, because at this stage of the game we're in pretty deep with them.

And cyclical links lead us to the point of the whole exercise; the superluminal aspects. How can we get around the cyclical scale speed c blanket? I know how superluminal aspects supposedly figure *within* sub-systems. What I'm asking is how do we get around superluminally within the stability of greater gravity? How do we traverse this or any other cycle in a manner that makes cosmological sense?

. *getting to the cyclical end-point*

So where's the reality in cyclical connections? I'm getting there. Like a tortoise maybe, but that's because we're still finding the way. You're not reading Einstein remember. You're not even reading the result of qualified cosmological analysis. To be frank, this is some unauthorized, unsupervised, self-styled Ph.d. thesis on international political science and philosophy. So figure that one out. All those years at university . . . Well, sure they wouldn't amount to a decade. But all of that Arabic and those Qur'anic translations, and this is how I think you can stop war-fare. It could be one way. The challenge is to unite for a common goal instead of in the face of a common enemy. And that's no big revelation. I've said it before. They've said it before, eons and eons ago, repeatedly down through the ages.

Cyclical links mean a beginning is also an ending of something else. They presume continuity within the cosmos, which our laws of physics don't. Or so it seems. And as far as we can see, neither does reality. Well, limitations and connections are what this theory is all about. How do we overcome them? How do we overcome the speed of light? The challenge is there. Cyclical links could play a major role in figuring all that out.

Already I'm suggesting that cyclical links would elucidate an order to events—all of them, right down the bifurcating line into the atom, as well as above and beyond gravitational expansion. So far, we don't get the greater framework that covers all that. What generates ongoing processes is apparently out of the picture. Symmetry however, presumes a certain pattern of events, cycles being one of them. I reckon they unfold automatically through symmetrical invariance. And the specifics would follow suit. So there are two issues here; the internal symmetrical processes, and the external ones. Your cyclical link is conceivably a product of both.

Look the universal expansion we're picking up could theoretically relate to symmetrical sub-systems, to some extent. Are we ballooning out then, just according to internal symmetrical manifestation? I hope not. But theoretically we're not simply subject to our hypothetical neighbour's situation either. Not that is, according to the grand-scale scheme of things. We shouldn't simply balloon out as they cave in. Although, I'm wondering that this macro turn of events could play a more decisive role than any internal mechanics. Even so, these internal processes generate mass and energy levels. And mass and energy levels give rise to systematic changes.

On that basis we should be able to tie the larger pattern of events in with system derivatives. How many leaves and branches does it take to amount to a tree? How many trees in the forest? I mean a tree isn't just a random event. Strip all the branches off and it's malfunctioning. And the big white wallaby, the one I've protected—Sooty's mate, starts looking pretty smug. So smug it's going for all the flowering budding varieties. Or so it seems. My cherry tree is stripped to buggery. They won't learn. Go on, live and learn wallos or you'll be pushing up my apple orchard. Anyway, these internal symmetrical occurrences should be relevant to the bigger picture. Each symmetrical event should involve force formations. It should entail a *systematic* transformation process. Your bifurcating levels have their own staggered force set-up. They progressively manifest decisive energy and matter ratios within spacetime and motion. And these specific energy and matter ratios trigger off a different scale of symmetry—a different scale of force formations. We have, for instance, the large scale pattern involving gravity's general relativity. We also have stellar, atomic and galactic ones. I think they all link in. And we should be able to reason them according to their symmetrical relativity.

Let's kick off with the cycle formation. Let's kick off with a cold one and the footy grand finals. Now that's more like it. Great isn't it—the way it lingers on long after the glorious event. At least up here in the hills it does for me. So back to reality Gloria. Back to today. There it all is out there. The grey mists rising up through the mountains encircling the verdant green valley. The fine rain gently falling and bloody Freddo croaking in the pond. I'll get him. I'll shove him into a cyclical end-point. Frigging frogs. In forming our cycle we should have three other forces. We should have four *cyclical scale* force formations all up. That means we have four different curvatures of spacetime and motion. Gravity isn't the only one. Gravity and orbital expansion wouldn't be the only determining features of a cycle formation. At some stage, for instance, a spherical spin could be the primary defining one. Righto then.

Righto then froggies, that's it. BIG fish are going into those ponds. So croak away my little cherubs, or hop back down to your thousands of rellies in the dam, because up here other company's coming your way.

Technically, we start with the hot energy of the big bang. That could be your large-scale spherical spin—force formation number one. The spherical spin, supposedly, has a combination of minimal spacetime with maximum motion. It seems to signify both the beginning and the ending of a cyclical formation. Furthermore, technically big bang energy transforms into matter as it cools down and expands. Energy might fall, mass might rise and cyclical expansion happens.

Theoretically then, this could be where we have primal bifurcation. Here's out internal systematic force differentiation happening. We've got rapid fire energy and mass changes. All of which, continue forth on slower levels. Change here is begetting change. Certain ratios of energy and matter lead to different scales of force formations. And at some stage they should all amount to a structural mechanism. That would be our living symmetry right down the cyclical line.

Now because we're getting this invariant change manifesting systematically throughout a cyclical scale force formation, we should be seeing *decisive levels* of transformation. And we should be on a multifarious journey with it. Your slower progression produces more intricate set-ups. Each decisive level should describe increasing networks. Take another tree, for instance. Look at it branching out from its trunk. O.K., so that's arboreal. But most things figure arboreally after a while here. They're everywhere around this valley. I've got mountains of them staring me in the face. But that's just for now. No doubt someone will want to transmogrify the lot into money one day. Anyway this internal business shouldn't just add up to a structural necessity of cyclical formation. It should serve another purpose. Staggered symmetrical invariance allows for element formation to occur systematically *within* the cycle. It's automating it. You're supposedly getting a fully ordered set-up inside as well as outside here.

Atomic systems for example, go into planetary solar systems. Then we have galaxies. Each system is supposedly subject to a decisive transformation process. Each one theoretically, should derive through the invariance of symmetry. I'm just saying that our symmetry is a multifarious beasty. Internally it seems to form a structure with many parts of great variety. Our solar system, for example, would be the manifestation of internal symmetry, as well as a consequence of cyclical events. And the whole thing gels together. All the sub-systems, all of the symmetrical deviations and derivatives balance out. Over-riding the entire process, at this stage of our cyclical life, we've got the general relativity of gravity. It's our proverbial tree trunk. The effects of it, whilst remaining true to its spacetime and motion combination, reduce as we get deeper into the sub-systems. Your leaf, beholden to a trunk, can still flutter around on it's branch. And inside your atomic shell superluminal party games could be in full swing. But all things being relative, this greater gravitational force seems to settle us into an evolutionary position within the cyclical systematic process.

Already I'm suggesting the greater gravitational limits are intergalactic. Inter-galactic mass could amount to a decisive dimension. You see; I can't make sense of endless expansion. There's no rhyme or reason to it. There's no balance to it. But you need the mass density of an *observable* universe to trigger a contraction stage. Or so it seems. Well, I'm factoring in symmetrical necessities. We're talking balance and formation here, not just what you can get through our best telescope. An expansion or a contraction phase should also rely on symmetrical invariance. I'm suggesting the cyclical formation is a contingent set-up. It's a matter of adjoining ones. Your expansion or contraction phase has circumstantial possibilities. It depends upon next door's situation. Their dark matter comes into play, one way or another. They could, quite likely, stop you from expanding endlessly or contracting inconceivably.

Look, this is where philosophy and science both make sense. You can even draw on Jesus here. Philosophically speaking, I reckon you're always subject to something else, so you might as well get to know it respectfully. Everything interconnects necessarily. Now I'm not denying either fatalistic or existentialist realities. I guess I'm just going along with Newton's science of causal phenomena. In that sense, the mass density of any one cycle would not *singularly* activate a contraction phase.

All things being equal, our grand scale gravitational orbital should evolve from beyond as well as from within the cycle. According to this scenario it also derives through the general relativity of force formations. The greater skeleton of which, defining the perimeters, is a give and take plural situation. Quasar activity could possibly be the out posts. Your quasar could be another cyclical scale force formation. Force formation number three of four—depends which way you look at it. I'm wondering it's a wave factor. Perhaps it follows hot on the heels of a cyclical orbital. Or maybe it precedes one. Whatever. The thing is; one cycle's mass ratio wouldn't directly activate the systematic *predominance* of another force formation. Rather that should also depend on the greater *cyclical* symmetry. And that also involves the formations of spirals, spin and waves.

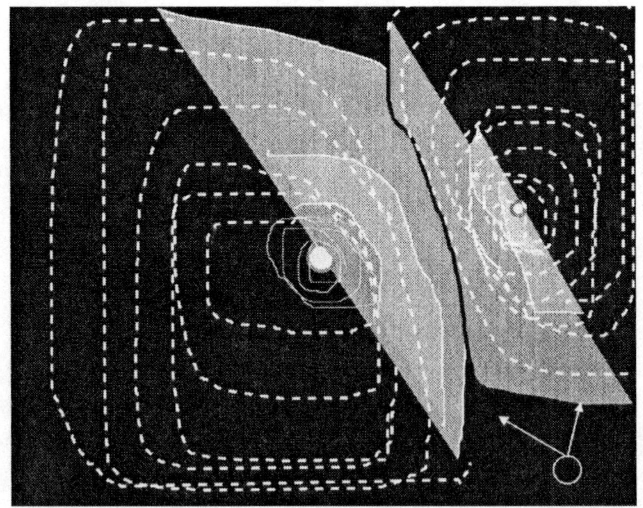

The above diagram shows some symmetry of two adjoining cycles. They both have an internal structure of decisive mass transformation levels. But they are two distinct set-ups, within a common broader system. We're looking at the one orbital formation via two cycles. I know it's not exactly your perfect round ring. Maybe from a different angle? What I'm getting it however, is that the expansion of one cycle would affect the contraction of another, thereby producing the shared orbital pattern.

This being the case you wouldn't get greater and greater conglomerations of stuff. A mega galactic cluster, for instance, might suit your expansion. So you've got your super clusters going into mega compositions. But a mega-mega is off the drawing board here. Cyclical contraction according to a neighbouring cycle would draw the corporate line on them. I reckon the gravitational force and its dark matter presumption reflects a cosmic order. One that involves expansion *and contraction* phases of cyclical dimensions. And there should be plenty of them.

Even so, we're still moving in the land of make believe with this, to a very large degree. You wouldn't get this far without infinity and eternity. That divine dimension transposes the finite boundary of spacetime and motion. Rather, the 'everything out of nothing' scenario is the stuff that fits into our observational abilities. I'm just trying to remind you these powers presently have limitations. And just because we can't see infinity and eternity doesn't mean we don't get its effects. We can still be subject to the causal phenomena of it.

You'd reckon the cumulative gravitational force of all the galaxies would halt cosmic expansion. What's cumulative gravity though? How many galaxies does it take? My position is that gravity has a formation. And this formation, on the macro scale, is subject to three other ones. What halts cosmic expansion could also be a matter of how the big orbital balances out. And that would depend on how other force formations figure. Well, Sooty my love, galaxies have decisive shapes. They're spiral, elliptical and spherical. And spiral galaxies seem far more common the farther out—and back, we look. And I know you're only in here with me because it's wild and wet out there today. I know you'd rather be up there in the bracken, under the fox gloves, serenaded by the little blue wrens—pouncing on them as it pleases you—sauntering down here for some tasty tiddy . . . hope the eagles don't get you.

Perhaps these galactic formations have a *forceful* effect. Perhaps they figure according to the symmetrical bifurcation that occurs within the general relative situation, which, in this case, is according to the macro gravitational force formation. Rhythmical, isn't it? And if so, their interactions are also a systematic element of bifurcation. They could figure more generally according to their rung on the bifurcating ladder. That's symmetry personified, lady. That's Ken banging the book-shelves into the wall. Well, maybe we could pick up where we are in the big orbital—expanding or otherwise, by the motions of them.

Besides, galaxies apparently evolve through 'density perturbations.' And that seems to concern the attraction of their neighbouring systems. Indeed, they could be on the cutting edge of cyclical internal and external factors. They could directly pertain to the inner structural symmetry as well as the mechanics of the large scale force formation. I just don't know how many it takes—your cluster or your mega cluster. Even the force formations of one galactic system could have a tangible effect on the cyclical general relativity. Your galactic changes could have such a strong ripple down effect upon all its sub-systems, that en masse for sure they're churning up dark matter.

"Apparently the arms of a spiral (galaxy) are due to pressure waves which sweep round the system at a rate different from that of individual stars. The added pressure in these waves triggers off star formation; the most massive stars evolve quickly and explode as supernovae, while the pressure wave sweeps on and leave the original spiral arm to disperse. If this is correct, it follows that no particular spiral arm can be a permanent feature. n b 1

Now take this one step further up the bifurcating ladder. Remember, as we go up it, our forces don't act with so much tandem gusto. The bigger the system the more independently they show up. Your spherical spin mightn't fire alongside its spiral, wave and orbital counterparts in the galactic system the way it would inside an atom. And by the time we're at the galactic level, not only is any one formation functioning more freely as is, it's also liberally reaching into the next level. Galactic spiral formations could draw into the spherical spinning density of a general level (cyclical scale) big bang.

Should independent spiral formations be increasingly prominent within a said cosmic area, then everything in it could accelerate. They've got more motion and less spacetime than waves and orbitals. En masse, they could trigger a general contraction of spacetime into motion. Well, in so far as all the other aforementioned factors allow. One such contraction might realise enough density of spacetime and motion to form a *macro* spherical spinning shape. This is where we have our generalised minimal spacetime and maximum motion set-up. You're no longer simply galactic, but in the cyclical turn of events now. We've come full circle to macro force formation number 1 again.

There's your neighbouring cycles drawing together into one spherical spin. The pink cycle is more prominent and closer to us. Even so, they should still affect each other. The black stuff is the dark matter. Without it you don't go in or out. It fills in the symmetrical gaps.

. the links

I think from here in on things should become more interesting. For one thing, we're delving into the superluminal arena now. Mind you—*symmetry, nonlinear, cyclical, superluminal, force formations, and bifurcation*; these terms are really getting a weathering. But you know, they're the basis of this theory. I can't get away from them. Take them as the tools of an unorthodox trade. Heavens any superluminal thinking has got to be beyond orthodox presumption, anyway. The orthodoxies probably wouldn't go near it with a 10 foot pole. Perhaps that too is a major hurdle for inter-galactic exploration. If no one writes about the possibilities for feasable inter-galactic travel, and no one reads about it, then how's it going to happen? Do we leave it up to the sci-fi worm hole experts and Dr.Spock? I mean it's right off the level agenda, imaginary or otherwise. So imagine that then, dooming ourselves to the destiny of this solar system through human complacence. Or worse, imagine dooming humanity to an implosion of human realities. We could remain subject to a crescendo of human incompetence. Either way, waiting around for a stellar superior to enlighten us without some serious self input strikes me as dangerous.* How do people behave in the yawning gulf of blind faith that someone will save them? Constructive past times? You must be joking. And what about those assertive independents at large? Testing nuclear arsenal; ripping through another rain forest or ten; breeding beyond consideration; all seem to be tempting fate—our own.* It seems we killed the last one, anyway.

So you see, I can feel quite noble in any forthcoming presumption. Black holes, quasars and accretion discs could be the mechanisms of cyclical end-points. The black hole could be the final contraction of the cycle, the accretion disc—an instrument of that. And the quasar could transport stuff into a new cycle. The quasar could be a macro cyclical mechanism emerging from a macro spherical spinning formation. And they should all have superluminal implications.

Now the last system up the bifurcating ladder—that I can judge, is the galactic one. What galaxies get up to, could directly takes us to the cyclical transition stage. Apparently they have cores acting like a generator within a powerful electromagnetic field. Theoretically, the core would be the system's spin. Or the galactic strong force. Arising from this spherical spinning formation we should also have waves and spirals. So perhaps our colossal electromagnetic field involves waves and spirals. It could all amount to a black hole, an accretion disc, and a forming quasar. It could also amount to Markarian 315, or something like it. n b2 The point however is; the galactic system, at least its symmetrical set-up, should resemble some aspects of a sub-atomic process. And the impact of them collectively should effect a larger symmetrical turn of events, the cyclical one.

Cast your mind back through theory. Supposedly, sub-atomic systems require balance, hence symmetrical invariance. Your nucleus would be subject to a given level of intensity relevant to the atomic *system*. When it surpasses that level it releases spirals. Now the same symmetrical imperatives should apply on the macro scale. Give or take, that is, the bifurcating ramifications of systematic hierarchy. At the galactic level we should still be getting the same 4 force formations. That's despite their varying rates of interactions. They should still be subject to the same intensities of spacetime and motion. They should still be up for symmetrical invariance.

Theoretically, symmetry functions through energy and mass transformations. And for this reason we have four force formations ranging in spacetime and motion combinations up to speed c squared. Now that's on the atomic level. Well, so too the macro one. If so, a dynamo at the core of a galaxy could be releasing spiralised energy according to the superluminal velocities of a great big spherical spinning formation—that's according to a macro symmetrical imperative. Revitalised with theory then, let's forge ahead into further presumption. The weather's definitely lending itself to it—briefly sunny, generally windy, definately cold, highly tempestuous and typically Tasmanian. A fish wouldn't rise in it. Far better being here near the fire then, locked into the computer for the duration . . . or making a plum pie hmm.

Our laws of physics don't seem to show that—not the weather, a galactic macro set-up of 4 force formations. I mean you don't get greater symmetrical imperatives with them. They only describe black holes according to mass, angular momentum and electric charge, *below speed c*. And that could amount to a linear framework, or what is only applicable to the spacetime and motion limits of gravity. And once again that seems to impose a finite aspect to spacetime and motion. So you wouldn't climb up the final rung of the bifurcating ladder with them. You wouldn't fathom a greater transitional change according to inter-galactic activity.

Using your eternal and infinite cosmic logic, however, things needn't pan out that way. With endless spacetime and *motion*, you're always transforming, for starters. There's always something going on. Things don't simply disappear down a black hole. And stuff doesn't become meaningless at speed c plus. This way, black holes just show the limits of the *general* relativity *of gravity*. The general level I'm taking as the cyclical one. That's the skeletal force formation level—the whole bifurcating ladder level. And gravity, with its specific combination of spacetime and motion, is subject to the systematic symmetry of this level.

Gravity, general or otherwise doesn't exist without reference to three other forces. The *general* level of it should also involve their greater balance. That means three other general types of relativity. And being general and of such a magnitude they'd probably predominate individually. I don't see how they could share the lime light as the sub-system forces do. Generally, I don't see how they could fire away together. But get this. They might not appear as general as each other. Their diminishing values of spacetime and increasing values of motion give them each a different dimensional reach in cosmic spacetime and motion. I mean your galactic black hole can fire away in *minimal* spacetime with maximum motion. It's a flash in the pan compared to your inter-galactic orbital, which factors through *maximum* spacetime with minimal motion. But on the cyclical scale one formation is balancing out another. One cycle's contraction should contribute to another's expansion. This is how these formations seem to reach across the system into the cyclical pattern. Put one formation along side another one, and you're looking at a set of cycles. Now I've just got to work out how they link in and define our cyclical junctions.

. *black holes*

Theoretically, everything has spin. Everything is subject to some curvature of spacetime and motion. Every force formation involves rotational motion. It's just that sometimes it figures as spin. Whilst at others, it could figure as an incremental crawling curvature. A star, for instance, can radically collapse in size. In so doing, it spins much faster and more energetically. This spin can even apparently produce more energy than nuclear reactions. And something like a 'flywheel' can contain it.

Well, you might think it all originates from gravity. Supposedly it's happening under the general level of it, anyway. The system sure is. It's still under the guise of the incremental general curvature of spacetime and motion. But here inside the stellar level, I'm thinking again. As you go up the bifurcating ladder these systems are more powerful. Greater is their ability to tease the general confines. With increased spin and reduced spacetime, such energy could be functioning beyond greater gravity's reach. And your flywheel analogy might be the stellar obeisance to greater gravity's presence. At the stellar level, a solar spin wouldn't have enough power to over-rule the tandem interactions of the other forces of its symmetrical scale, nor black-hole into a general big bang situation. Still, a collapsing star could be entering a different force formation situation—one that factors according to a different curvature of spacetime and motion. And one that has a different combination of spacetime and motion, albeit within the symmetry of its own scale.

These star changes could mark the beginning of discernible superluminal activity. The 'flywheel' seems to signify a potential challenge to the speed c limit. It seems to maintain greater gravitational stability. It could contain the presumption of a new force. Here we have greater energy and spin, instead of greater spacetime. I think an inordinate increase in matter density triggers this. It produces a shift in the spacetime and motion curvature. Your spin is chewing up the spacetime. Your solar spin, I think has the ramifications of a black hole according to it's system. Look your black hole ostensibly happens via a matter *density* increase. And it goes beyond gravity. There's your radical concentration of mass that could be effecting a forceful combination of spacetime and motion. How all encompassing the spin becomes, as in a black hole, probably depends on its scale. The galactic one, for instance, should be more omnipotent. Well, O.K. then, not exactly omnipotent but on that road.

We can look it through another angle. Theoretically, an increase of spin amounts to greater nonlinearity. And that can fuse mass with motion. The mass-motion fusion functions with less spacetime. But here's the point. Mass-motion fusion means superluminal activity.

Our laws of physics don't look at things like that. Sure, they point towards black holes. Heavens, general relativity predicts their existence. But they do so with their limitations. That's as far as you go. Well, that's as far as the general relativity of gravity goes. That's as far as you go below speed c. So enter the black hole. It features all the questionable stuff at the edges and beyond the known laws of physics. Stuff like quasars, superluminal forces, strange stellar activity—the black hole seems to cover the lot.

I'm agreeing that at the black hole stage of events gravity is questionable. How questionable though, could depend on the scale of the system in which it's arising. Yet whatever the level, I think we're going into a spin situation that according to our laws of physics only factors in the micro world of quantum theory. When there is an imbalance of matter and energy, the symmetry of spacetime and motion should act. When matter and energy churn up enough spinning motion another force comes into play. It could eventually affect the general predominance of the gravitational formation. Another force could generally predominate. But going by the stellar scenario above, it's a staggered situation. I don't see one *stellar* black hole ousting greater gravity.

If a galaxy has one we call it active. And now we could be talking general effects. Now we're going into the land of quasars and accretion discs. I reckon a black hole and a quasar can form a galactic engine. And there aren't many around here to say otherwise. Evidently this galaxy is pretty well stable, being in the throws of a gravitational general relativity. At some stage though, these engines could put the wind up the sails of the greater orbital formation.

Be that as it may, the scenario is riddled with loop holes. For one, according to our cyclical level these formations should figure one after the other rather than in tandem. They should feature consecutively within the singular cycle. Galactic quasar, black hole and accretion disc could, however, amount to spherical spin, waves and spirals factoring in staggered unison, rather than individually predominating. It's possibly more of a team effort than the one-on-one situation the general level assumes. So I've got these things figuring systematically in galaxies, yet singularly featuring inter-galactically.

Makes me wonder that a nifty nonlinear equation wouldn't go astray here. If x equals y then b follows automatically. One way out of this is the misnomer of a singular cyclical set-up. Theoretically, there's no such animal. These general formations have to figure alongside each other to make cyclical sense. You don't get one on its own. They define each other. And on the galactic scale we're supposedly still one rung below the general top of the bifurcating ladder, anyway. So at this stage these formations logically still figure more-or-less together. It's only when we reason through their inter-galactic effects that we get the general picture. That's when we might be getting the general big-bang black hole, or the general wave or spiral effect. In any case, a greater black hole situation should contribute to a cyclical end-point. The massive big bang that kicks off an expansion phase should out perform any flywheel addendum.

In other words, a general level black hole contributes to a process whereby energy will start forming matter. We're drawing towards a cyclical junction stage with matter contracting to the degree of mc squared spin. That's the predominance of a strong force formation on the general cyclical level. That's your $mc2=E$ on the cyclical scale. And when it's E then it can start becoming m again. It can go into spirals, waves, orbitals. More of that later though. That's another cyclical expansion.

. *accretion discs*

Although an accretion disc apparently figures with a binary star system, I think they could have a bifurcating counterpart acting as a galactic mechanism. Take a galactic engine. It could comprise of a big black hole with an accreting set-up whirling around it, funneling matter inwards. Ostensibly that's our spherical spin and spiral formations. These two formations seem to power systems along nuclear lines. And they could churn stuff out again through waves into the gravitational stability of an orbiting galaxy. To date however, any such galactic engines probably register as the macro quandaries of sub-atomic uncertainty. When it comes to spirals and spinning formations, no matter what level, it seems we're still up against the superluminal gap in our intelligence.

Despite that, accretion discs could have more going for them than black holes. Their *existence* seems to point to the *likelihood* of a black hole. Theoretically they pave the way for them. The logistics of force formations leads to them. I mean your spiral should factor alongside your spin. You wouldn't go directly from orbital into spin. You wouldn't just have an orbital going into a black hole. These force formations should transform systematically. The orbital stuff should *spiral* into a spin. Taking general relativity further than gravity allows, reveals a systematic process of events. On the general level we should have a symmetrical transfiguration of the gravitational force formation. There should be some spiralised set-up that's instrumental here. For this reason, I'm going with accretion discs.

Too bad they only show up on the stellar level, and that's by way of binary stars. What are their bifurcating possibilities, then? How do they fit in generally? On the stellar level they seem to relate to the flywheel scenario. Perhaps their realisation is the physical evidence of gravity's systematic limit. And perhaps they act for the same flywheel reasons. They too appear to be a product of symmetrical invariance. That is; an accretion disc should maintain symmetrical stability via greater gravity's presence. Now galactically you still get that. We remain under the guise of general relativity inside the galaxy.

Still on the stellar level, the *binary* system shows the gravitational forces drawing in the matter of its companion star. That's your accretion disc forming. That means one star's mass alone is apparently insufficient to generate enough gravitational intensity. You're not straining the system with it. You need more than one sun for it to figure. Because it's drawing from the stellar system's gravitational limits it's probably rebounding into the galactic level. Now take the logic one step farther. We're arriving at the general rung on the bifurcating ladder. With two or more galaxies we're talking inter-galactic activity. We're considering the pressure of adjoining galactic systems. Combined, they might cause a general contraction of spacetime and motion. That is; inter-galactically, the contraction into one such accreting formation might be great enough to out weigh the general gravitational stability.

On the stellar level the spiral-spherical spin relationship forms stars. Your accretion disc-come-black hole is possibly taking the stellar system to the next level. It could be defining stars according to a galactic system. Inter-galactically however, an accreting set-up could manifest enough nonlinear curvature of mass and motion to herald in a big bang. A greater accreting set-up could over-power the presence of other force formations, completely. It could singularly predominate over a whole bifurcating structure—the way gravity does. That means our accretion disc could generate enough spherical spin, or strong force activity to start a new cyclical phase. We're coming full circle with it. When the sphere is too intense, just like a sub-atomic charge system, it should release nonlinear mass-motion according to waves. (And according to your neighbouring cyclical set-up, and its preponderance of dark matter.)

Below is one heating up in the middle. There's your greater spin. It's increasing spacetime compression and nonlinear motion at its core—hence, the spherical spin of a black hole. When that becomes too intense it gives off the macro version of radiation.

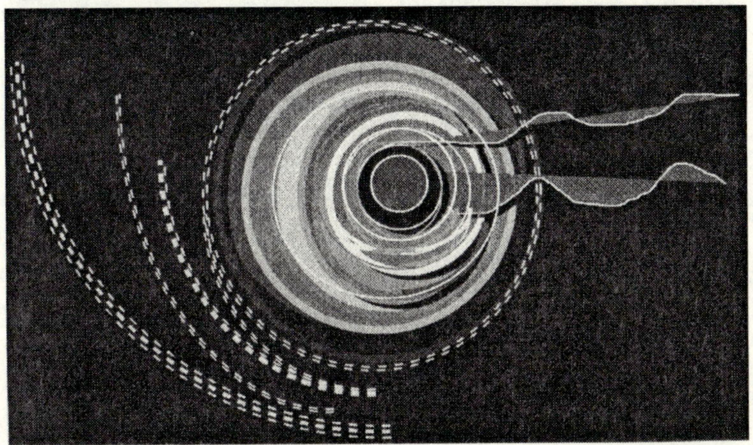

Technically, the stuff in the inner and outer parts of this formation manifests friction. I reckon what goes into one can black hole and then come out of its hot inner edge as radiation. Theoretically, that's when the disc turns into an enormous dynamo. The black hole spherical spin is generating electromagnetic fields. Galactically, we've now got stuff expanding outwards from a 'nucleus.' And I think we're talking quasars at this stage. Perhaps this is where we have our star bursts and cosmic ray activity. Perhaps that's Cygnus x-1. Below is one such dynamo. We've got stuff spiraling into the central black hole of an accretion disc and transforming out of it as an electromagnetic release.

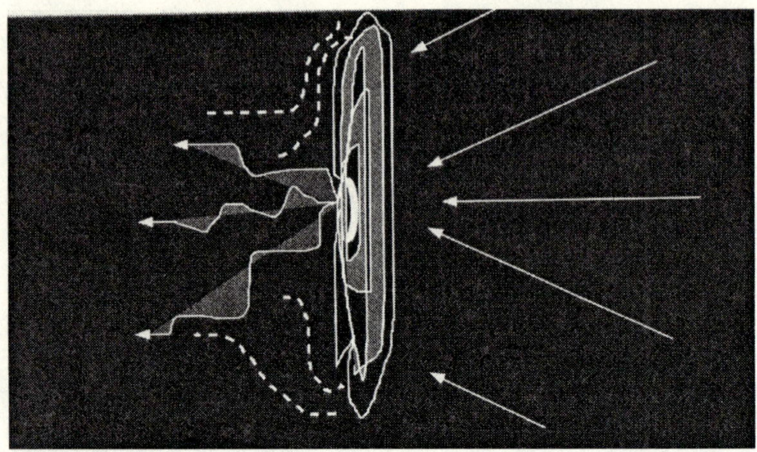

If nothing else, the greater accretion disc might dispel images of a black hole drainage system in a galactic sink. You see; it doesn't just arise through intensifying gravitational power. It also seems to process it. Stuff could transform and resurface by way of an accretion disc rather than disappear forever down the black hole gurgler. Your accretion disc is paving the way for ongoing transformation processes. And on the general level, this spiral set-up should effect the predominance of a new force. Indeed, it should itself predominate for a while.

going into quasars

How much power derives through stellar accretion into a black hole? Enough apparently, to spark a quasar and empower a galaxy. The quasar I think, is transforming energy and mass into galactic proportions. Its massive radiation can outshine 100 billion stars. There's your new galactic core. And the logic of it translates to the general chain of force formation events, size-wise particularly. Quasar radiation makes for outbursts through a source the distance of our solar system. One such formation can generate the equivalent power of illuminating 10,000 galaxies in 1 day. Now that's some mystery our laws of physics can't explain. I expect that's a galactic one. But illuminating 10,000 galaxies in one day? Wouldn't that relate to the inter-galactic general level?

Inter-galactically, the quasar could be expressing the general relativity of the electromagnetic force. That would be the first step into a new cycle. It could be establishing the rungs on another bifurcating ladder. Technically we're going to work our way from E back into the mc squared of another cyclical scale big-bang spherical spin. With the macro wave formation we're drawing out of the condensed spherical spin matter of mc2 that has big-banged into energy.n b3 Systematically, we're now proceeding into another phase of greater gravitational stability with it. Our general energy is once more expanding and stabilising into matter. You don't normally see it that way though. The big wave and the big spiral doesn't normally figure in the general logic. Only the big bang and then a gravitational spacetime curvature does, and within that atoms start forming, etc. Sure, greater gravity may assume the dominating feature of the system. But that, I suggest is due to its maximum time and space quota with minimal motion. I mean gravity figures for a longer time over a greater distance. Against the expansion and contraction of a general orbital formation, the rest of these generalised force formations wouldn't appear so great. Their bigger hit of motion makes them shorter-lived. All up though, we're now theoretically reckoning the systematic symmetry of a cycle in full swing.

Let's do it again. Yes, maybe more beer with a modicum of restraint might be the go. It's a strange drinking technique this home brewer—I'm wondering—needs to come to grips with before proceeding with batch number three. The stuff clearly isn't worth the effort until it's over 2 months old. And if you're guzzling you're behind the 8 ball, all the way back to the jolly bottle shop. But putting beer aside, let's go through the four force formations again, on the cyclical level. Through the brewing inter-galactic pressure, we overcome the prevailing intensity of the general relativity of gravity by spiraling onto an accretion disc. Around we go, into the overwhelming strong force of the spherical spin of a black hole. Out we come, radiating in quasar waves. Clearly sober, there we are, in a new galactic expansion of greater gravity's orbital once more.

Think about it. The nearest known quasar is 2 billion light years from earth. Now the only galaxies that don't appear to be expanding away from us are the members of the Local Group. That's our locale. And the quasars seem to figure beyond that. They seem to be far enough away to factor as a denizen of cosmic expansion. But get this. The farther out we look—the more numerous they become. According to theory, we might be detecting evidence of cyclical junctions way back in time. The question is; do they signify the expansion phase of this, our most immediate cycle? Or are we detecting evidence of adjoining cyclical processes? Let me quote something:

'Halton Arp, formerly at the Mount Wilson Observatory and now working in Germany, has found that there are pairs and groups of objects (quasar/quasar, quasar/galaxy, galaxy/galaxy) which are connected by visible 'bridges' and must therefore be associated, but have completely different red shifts. If so, then the red shifts are not pure Doppler effects; there is an important non-velocity component as well, so that all our measurements of distance beyond our immediate neighbourhood are unreliable.' n b4

So the next question is; if we're really in a cosmos that contains a myriad of cyclical expansion and contraction phases, what's with the general expansion away from us that we're picking up? Have they all been to see Star Wars, parts 1 through to 4? Is our bubble mentality detectable beyond our luminal notice? Who knows. The thing is; quasars show us the possibility of superluminal activity, and their mechanics could spell out the secrets of superluminal travel. Without it, we're not going full circle into ongoing processes, theoretically or otherwise.

In short, the quasar could be the fourth and final link in the cyclical chain of events. It could set the scene for ongoing transformation processes on the greatest level. So what about cosmic rays, you may ask. Aren't they the big bang background radiation? Going by the above, your cosmic ray should come from a systematic process. Why should these things all stem from an omnipotent big bang? Theoretically, the big bang is just the focal point of an accretion and quasar association. Besides, technically the inner hotter part of an accretion disc (the area pertaining to the black hole-come-big bang) produces ultraviolet and x-ray photons over a broad range of energies. A cosmic ray might emerge from that lot. And the uniform nature we detect in them could just refer to the following prevalence of greater gravity.

. *cyclical links and micro counterparts*

I don't see that the macro is a neat replica of the micro. Although the symmetrical derivation of one from the other should mean there are decipherable correlations. Look, we've got our arboreal analogies. We've got our bifurcating ladder. Clearly on the cyclical scale we have less force formations. That's where four macro forces interact to form the perimeters of *cycles*. That's your skeletal cyclical framework. It's not the same as the atomic system. On the cyclical level we conceivably have a *primordial* differentiation of spacetime and motion.

Unlike the merry proliferation of atoms, we should have an endless ream of primordial differentiation. Their cyclical pattern, however, should form an outgoing myriad rather than an inward manifestation. Inaugural systems should endlessly extend through infinite and eternal spacetime and motion. But do they *conceivably* do that? Am I making any sense thus far? Or is the bit on not guzzling beer the only bit of verbiage worthwhile? Could be that's the only bullshit thus far. Either way, is it adding up yet? If the micro systems don't align with the macro ones logically, then I've lost the plot. We might as well stay with the uncertainty principle. Or continue fashioning links between an atomic nuclear spin and a greater gravitational expansion—with numbers. Our infinite and eternal spacetime and motion presupposes order based on symmetry. This cosmic prerequisite should extend from its primordial division into internal sub divisions, with no chaos in sight.

It's not just for logical validity that I want to draw micro/macro correlations. Although sure, that's in there. We're wrapping up the whole cosmos symmetrically and systematically here. If chaos still lurks in some corner of it, the rest of it might remain questionable. I'm advocating a bifurcating symmetry within a cyclical pattern. I mean you wouldn't get bifurcation across the board. Bifurcation stems from your cyclical links, which themselves form the primordial symmetrical pattern. On that basis, we wouldn't find micro counterparts to these cyclical links. Your atoms, for instance, shouldn't link in via a shared mini big-bang the way your general level might. They don't need a communal proton spin to form a set of atoms. But we should still be able to discern the systematic interaction of four force formations pertaining to different scales. And so I do. What's also becoming apparent is the smaller the level the more unified are the four force formations within their system. The mini electron orbital, for example, happens alongside its mini proton spin, and simultaneously manifests waves. All four micro force formations fire away and feed off each other, at once, to realise their atomic system. As you go up the ladder however, these force formations individually function more clearly. I reckon they're less dependent on each other for symmetrical expression. Indeed, their systematic realisation increasingly attunes to the general scale of symmetry. Until finally, one formation alone has enough power and presence to predominate generally. And when that happens, it's also defining an adjoining system. For balance it's now depending on the primordial differentiation. It's overcome the entire bifurcating ladder and is forming the cyclical pattern.

So there are two points about all of this. First, for symmetrical and systematic validity, our four force formations always operate according to an unchanging quota of spacetime and motion combination. A wave, for example, on any level has possibly equal amounts of spacetime and motion. I'm not sure of the exact proportion, but what I can safely gauge is that its proportions render the motion quota to speed c. That same combination figures in the atom, in the stellar system, on the general scale, and any other scale you care to define. With your orbital, though, you're always looking at minimal motion to maximum spacetime. And so it goes on every level. The second point involves the differences between forces on these staggered levels. They interact differently. On the bottom rung of the ladder, the atomic forces seem to interact simultaneously, thereby forming an atomic shell around them for social cohesion. As you progress up the scales they interact less in tandem and are more open to the greater symmetrical situation. That is; the general relativity of gravity wouldn't pervade the atomic shell they way it would inside the stellar system.

By the time sub-atomic systems function they're beholden to plenty of other symmetrical dimensions. No wonder they've got an atomic shell. Does a proton inside an atom on a planet tune into the whole cyclical symmetrical set-up, and primarily spin according to that? Sure your leaf can flutter around on its branch, but that's just it. It's not fluttering on the trunk. It goes via the branch and its little green world up there. And the branch tunes into the trunk. No leaf—so what. But no branch—no leaf. And no trunk—no nothing. Admittedly your tree would have a problem, if it's not a deciduous one, that is, without any jolly leaves at all. But now we're past the point of pleasant pedantry. Better climb out of the tree and get on with it. In varying strengths, the atom comes under the symmetry of a solar system, the galaxy, and the general situation. So you have to get into its shell to pick up on any superluminal velocities. And you should be able to do it along side the other three atomic size forces. They fire away together remember.

Take a step up the ladder though. On your stellar level, general relativity is probably more pervasive. And your forces wouldn't fire away so much in tandem. To pick up on any superluminal activity you probably have to get inside the relevant force formation, not just the system. You might need to get inside the spherical spinning set-up of the sun for instance to gauge the stellar variety of superluminal strong force spin. Take another step up. Galactically your black hole is even less in tandem with the forces of its system. And at that level it's approaching the measure of the general relativity. So the point is, the symmetry required to keep the forces in check shifts as you go up the ladder. The farther up you go the more the entire system links into an individual force's activity. The lower down the scale, the less impact either the individual force or the general symmetrical system has on each other. On that basis, I don't think you would generate an atomic nuclear spherical spin without its atomic coordinates. They would also shield the spin from the greater symmetrical persuasions. And at this rate, particle physics is seeming like a blessed relief. As a way of working things out, just give me a quark instead. And make it a double please, with zions.

Which brings us to the point of the points. This being: how to produce the superluminal power of a strong force formation. Theory has it that the spherical spinning formation always operates under the same combination of spacetime and motion. It follows that your strong force is always happening at speed c squared within minimal spacetime. Well, I can't see us tapping into a macro version of one to get off the ground superluminally. But in looking at the force through the progressive levels within a cyclical set-up, we could decide its interactive abilities and so power to act. The macro spherical spin, for example, should over-ride all subsequent replications. It's more powerful and durable. Its atomic micro counterpart, however, interacts more unanimously with the other three forces of it's micro system. And it superluminally does so under the guise of the general affects of the greater one. Its systematic set-up is allowing it to. Again, the general affects theoretically only concern the atomic shell.

Well, just how deep do its systematic abilities go? I know we've done the leaf analogy, but how much does the proton spin according to scale? The rate of force interaction staggers up the ladder. Plus the force has a symmetrical application. Everything does in this story. Your proton's actions should still contribute symmetrically, albeit intangibly, and quite likely—abstrusely—going by me, to another level. Anyway, it might need to think it's doing that to produce the power it does. There's a lot to consider then.

Although I'd dearly like to, what I'm getting at is we can't forego all the symmetrical nuances along the bifurcating ladder. And none of it could make sense if we don't determine any semblance of order. To understand all the ramifications of symmetry we could use correlative guidelines. It's a lot to work out though. And don't count on me. Who's the mathematical wiz or nuclear genius around here? Even if I was—and we both know otherwise—my talents better lend themselves to working on this strange new drinking technique on Saturday afternoons, even wet ones. You know the one relevant to home brewers—taste testing. Here then, remains the correlative issue for now, up for contemplation.

I suppose to clarify correlations it would be wise to start once more at the systematic beginning. The general scheme of things shows symmetry basically acting through a colossal system of forces which form cycles. Taken en masse, cycles seem to present a balanced basis for infinite and eternal spacetime and motion. They can go on and on, forever, and ever amen. A cycle though, should also have a singular inward systematic manifestation. One that still remains subject to this greater symmetry. We're talking orderly bifurcation here. Each one should set the boundaries for internal transformation processes. All things being basically simple, as well as systematic and symmetric, your cycle is theoretically augmenting and then replicating the same differentiation of spacetime and motion internally. Why rock the boat? It's sticking with what works.

Balance is begetting balance. If so, on a diminishing scale within each cycle, we should have increasing amounts of four force formation sets.

Now that wouldn't happen externally. The logic of this balancing act means you wouldn't get bigger and bigger cycles. Once you're on the level where four whopping formations individually prevail, you've pared down the system's approach to the primordial pattern. That's the cyclical perimeters of a plural business. And look, one cycle wouldn't form through gravity alone. And four formations alone wouldn't form one cycle. It makes no symmetrical sense. They form the pattern of several. Each cycle is theoretically subject to the external dynamics of another one. You can't have one without the other. Play it again, Sam. One wouldn't exist without an adjoining one. They share, as a system, the minimal situation of their symmetry. Their pattern forms the primordial differentiation of spacetime and motion. The cosmos happens. How many ways can we put this? Out of the cosmic one, comes four. The cosmos then, can invariably function through its own symmetry.

On that basis, it looks like a system of cycles only requires the same minimal set of four big force formations. This way, cyclically, infinity could remain just that, nothing more and nothing less. And so would eternity. We've got conservation of the status quo. However, what I'm getting at is these four general level force formations don't just effect a cyclical pattern relevant to an eternal and infinite cosmos. Spacetime and motion mean necessary change. And change can have a raison d'être besides preserving posterity. These formations could progressively bifurcate inwards within each cycle to maximise change.

We're getting value for money here. All of these bifurcating manifestations provide structural opportunities. We're getting the means for manifold varieties of phenomena. We're getting zillion's of possibilities for change. Nor are those voids there for nothing. Everything is necessary about this remember. I'm leaving nothing to chance and no room for chaos. Indeed, their inward manifestation should also maintain the status quo. That means the general level also derives, and remains subject to its bifurcating manifestation. Your macro stability still relies on its billions of millions of microcosmic replications, and all the stuff that realises.

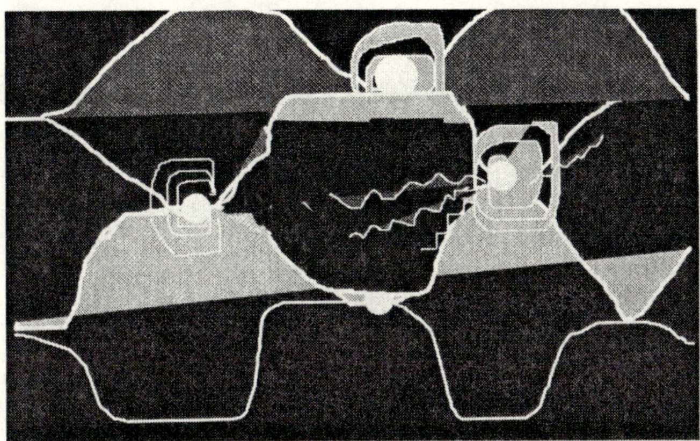

In this diagram we've got various cyclical stages. They're forming through the symmetry of four force formations. As above, these are: the dark blue gravitational orbitals, light blue waves of electromagnetism, and the nuclear forces of green spirals and yellow spinning spheres. And sure, that doesn't describe any inward bifurcation. But who's drawing that? You must be joking. I'm no Michaelangelo. Besides, the colour computer fortuitously, it now seems, caved in. As far as I know it's still in Launceston awaiting quotation approval. And unless money falls out of the sky—and I don't see *that* happening, I'm working with Harry here, the F.J. model in black and white. I only hope after 3 more months of this Harry's harmonics haven't shattered the molecular structure of my pearly whites. But getting back to the cyclical point, what I've got in mind is a tree with the branches, leaves, et. al., bifurcating inside the trunk—the trunk being the general prevailing force formation.

Failing the graphics then, I suppose I'll have to come up with a lucid account of every symmetrical manifestation of the jolly force formations per cycle. How to decide then, in a nut-shell; what are the correlations between a macro level of minimal formations, and their progressive microcosms of billions of millions? Where's that Sooty? Perhaps we should go for a walk down to the shed to check out the rabbit scene, and see if any brew's exploded. Perhaps we should proceed past there to where the big white wallaby hangs out. Perhaps there's gold in that little crickling creek in the dell— and here we go colour graphics again. Then again, 9 times out of 10 the rainbow lands on the dam up here . . . hody humm. Perhaps I should come to grips with the question and just get on with it.

Before getting into any staggered situations then, I want to go over the 2 most decisive levels: the atomic and the general (cyclical). Theoretically, there's a limit to the proliferation of diminishing force formations per cycle. With balance in all directions, at some stage one level holds the other in check. The millions of diminishing replicas secure the skeletal trunk. That means you can only proliferate inwards so far with this, and vice versa. It should also invalidate the butterfly effect. Symmetrically, your individual atom does bugger all here. En masse however, it's having a staggered effect. I mean if you want to evaluate the individual atom with the general scheme of things, then welcome to that wilderness of non aligned general relativity and quantum theory.

The purpose of symmetry, I think, is that it creates order. Symmetry appears to realise decisive invariance, and decisive levels of this for transformation purposes. Just like the supposed omega factor of expansion. On the cyclical level of basic formations, symmetry is replete. An inward manifestation of these formations should still be subject to that balance. They only take you manifestly so far within. I don't mean you can't dissect sub-atomic systems. Clearly we can. Dissect away. But I don't see any four force formation set happening within your zion or pion.

Furthermore, like most things it's probably an incremental process. Evidently, we don't go directly from atom into big-bang. Clearly we also have galactic set-ups and stellar set-ups. Just as evidently and clearly, we can't align the general relativity of gravity with the quantum theory of atomic nuclear forces. Ostensibly therefore, we might need to reason through the staggered levels inbetween. We might need to address the symmetrical ramifications of their links. And what a mish-mash of convolutions they could turn out to be without some set guidelines.

So in a pretentious attempt to keep things simple, I'm sticking with four levels. The main ones I can work out are the atomic, stellar, galactic and finally, the general cyclical levels. In all reality though, there's systems within systems here. We get the weather patterns of planetary systems, mountain ranges, trees, and countless collective items of balance with their own forceful impact. The list, digressively, goes on. Remember; we supposedly get the maximum potential for change with this bifurcating set-up. Systems evolve, expand and cool down. Energy and matter transforms accordingly. You live, you die, and transform. Theoretically such happenings go on forever in every conceivable form by and large, as well as in nooks and crannies through out cosmic wonderland. But for the sake of symmetrical cogency, they're neither here nor there. I'm forthwith only relating to atoms, stars, galaxies and cycles.

Next, let me re-state another basic guideline that applies on any level. Great isn't it. I'm so deep in theory now that we're getting guidelines. One can only hope they still relate to reality. Anyway, as variations of one force, each force forms a symmetrical element. So when one stresses its limitations another one could interact, all for the sake of symmetrical balance. When the spherical spin of a nucleus becomes too intense, for example, other formations step in. Our combination of spacetime and motion is pushing beyond a symmetrical limitation. It's inevitable. We've got motion factoring necessarily within spacetime in this theory. But that's not the only reason. Applying this guideline to different levels bears different fruit. Enter the bifurcating nuances. Thus far I've only alluded to them through the varying rates of force interaction.

As we proceed up the bifurcating ladder we're subject to manifold symmetry. And it becomes evident through energy and mass transformations and the direct impact our force has upon them. Again, these symmetrical directives are showing that sub-atomically our force formations are firing away practically instantaneously. The sub-atomic intensity of nuclear spin is *continually* straining atomic symmetry. So its 3 other mates are always actively around. It's all for 1 and 1 for all inside the atom. I don't think your spherical spin could foment nonlinear intensity all the atomic time without the balancing activity of spirals, waves and electron orbitals. How would the atom be otherwise? Theoretically then, no atomic-sized force would single handedly manifest a mass or energy imbalance. On this rung of the ladder it's got a host of symmetrical nuances weighing down upon it. Ostensibly, it's toeing their line, via its system.

In other words, your individual force here, only tunes in *systematically* to the bifurcating ladder. As such, your proton might primarily spin according to its atomic band of men. There's your tangible bifurcating impact. On this level, your individual force only figures according to its system. And your atomic spin can mean bugger all to the general relativity. So maybe the tiny one doesn't consider mass and energy the way the macro one does. Its system is a tightly knit affair, bonded together by symmetrical over-heads. According to this scenario, a singular micro force would have little bearing on the overall situation of energy and mass transformations, whereas one on a greater level might. So theoretical what, do I hear you say? Well, O.K. then; I'm getting there.

It's no leap of faith to see that each level of force formations should realise different mass and energy transformations. Nor that an over abundance of either can imbalance the system. The point is; their systematic rate of interactions is not just a decisive factor here. We're now considering how the force factors according to energy and mass ratios. These are another symmetrical imperative subject to the bifurcating ladder level. Too much mass on your stellar level and you get a greater gravitational pull, for instance. Then your stellar accretion can kick in and step the whole system up to your galactic framework. Now I can't see that happening on the atomic level. I don't think your micro spiral would kick in *after* a gravitational imbalance of mass occurred inside the atom and single handedly transmogrify the entire atomic set-up onto the next level.

Even so, we shouldn't just be looking at a systematic rate of interaction to understand our force. You see; the spherical spinning force is the one I'm targeting to get us off the ground. If we're going to churn out superluminal propulsion via the atomic version, we're obviously going to consider energy and mass factors. That's what the force is all about anyway: matter/mass and energy transformations. The 'force formation' is, after all, a manifestation of energy and mass. Who sees jolly gravity? You see it's physical effect. Amazing isn't it—coming up with such profound platitudes so far into the text. What I've been getting at though, are symmetrical and systematic directives. Theoretically, energy and mass symmetrically pan out on levels. And the levels show force formation systems. All of which bear allegiance to the greater symmetrical good of the cycle. We don't want our excess superluminal energy systematically swinging into the next level according to the general relativity. We want to circumvent any bifurcating strings of symmetrical insistence.

And what happens next with all of that, I'm not sure. But I can apply the logic to the contraction stage of a cycle. Take the stellar level of force formations again. Stellar collapses might cause sufficient energy and mass transformations to draw into a different level of force formations. In contrast to its micro counterpart; a stellar force formation could more readily react to energy and mass levels and change the system. With several of them you're probably looking at a major symmetrical shift. These bigger force formations have a potential for a broader, more far reaching impact on energy and mass ratios. One reason is because their macro *spacetime and motion combinations* can cover and drag in internal bifurcating sub-systems. Furthermore, they're theoretically not acting in tandem as the smaller counterparts do. On that basis, they're not continually in check. One could obtain sufficient stimulus to directly affect the stellar balancing act. Several of them might overwhelm the whole stellar level, registering an energy and mass imbalance on the next level. If so, your galactic formation is up for a symmetrical shift. Which means we're now dealing with galactic sized force formations.

What if an adjoining galaxy is also going through stellar collapses? We could be heading towards a cyclical end-point at the general level. We're talking big shifts on the ladder now. There's your symmetrical domino effect. There's metaphor number zillion and three. So what. The point is, instead of manifesting new levels of force formations, we're chewing them up until only the macro one is functioning, say of a spiral accretion disc. Unlike my bum, everything in the definitive area would be shrinking into a tighter non-linear situation. More intense mass, more intense gravity to figure in less spacetime and more motion, which means whammo—down the black hole gurgler we go. Mc2=E. So you've got more room to move next door. On the other side of the cyclical fence the stellar ones could still be forming. Galaxies could be spiraling out. And so it goes.

You can look at it from other angles. With a major symmetrical shift on the greater levels, we're challenging the general formation. You've got a more direct link here. The generally predominant force (e.g. gravity) has a deeper influence *within* these larger systems. It's suffusing the stellar forces' individual abilities to generate matter and energy more so than the atomic ones. The general formation pervades the system, (and thus its energy and mass ratios), the larger and more open the system is. Whereas the deeper in we go systematically, the more energy and matter transformations respond to the bifurcating balances and checks. Stuff, for example, wouldn't function within a nebulous galactic formation as it would on a planet. The stuff in a nebula would be subject to a larger, less intricate array of symmetry.

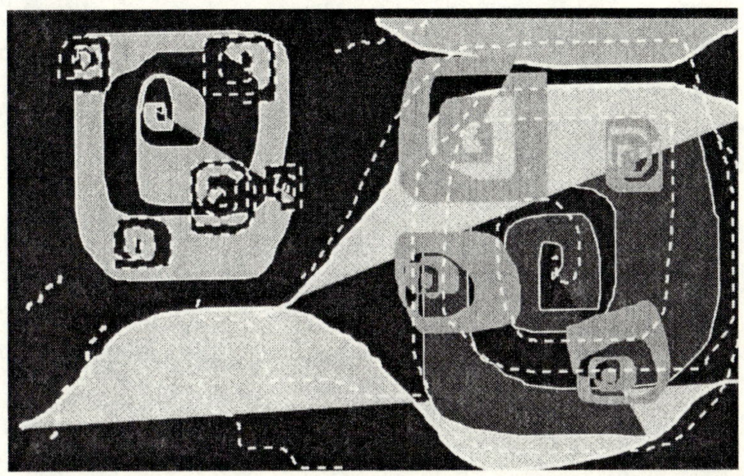

In the above diagram are some vague, largely theoretical, and unedited systems within cycles. The pink system on the left is contracting. The green one on the right is expanding. And the blue lines define the cyclical scale symmetry of both. The more the pink one contracts the more spacetime and motion there is for the green one to carry on evolving. Oh, and the staggered levels of symmetry within each could be stellar and galactic.

Well then, what of the central issue: micro/macro correlations? Where are they? And what's with all the bifurcating nuances? Without them, we'd probably have neat replicas rather than systematic correlations. Theoretically the energy and mass ratios carry the consequences of force formation interactions according to the bifurcating level. That is, whilst the spacetime and motion combination of the force formation supposedly remains invariable—despite bifurcation, energy and mass ratios relative to the force don't. Sub-atomic particles, for example, could be on the veritable cutting edge of mass and energy transformation and force formation. We're closing the gap between cause and effect here. We've reached the point where we can't bifurcate any further. At least not as far as I can tell by way of a force formation system. There's no room to move baby. You'd think that wouldn't you. And maybe there isn't. For energy and matter to count according to these four forces maybe you need a minimal amount of spacetime and motion *per se*. Maybe you need your micro *dark matter*. But that's not my reasoning here. I reckon we're at the bottom rung of bifurcation according to the greater symmetrical logistics of the entire ladder.

The sub-atomic lines of force seem to merge the formation within the mass and energy transformation. On the micro level, you've got a constant struggle for position. Take, for example, the sub-atomic action between electromagnetic wave formation and particle positioning according to gravitational formation. Whereas on the stellar level, it's easier to distinguish a causal process between force formation and mass/energy transformation. Planetary mass attracts gravity. And it goes into orbit. Plus on this level, your superluminal spherical spin could neatly wrap up inside a solar formation along the orbital by-ways some. Go back to your micro correlation though, and we've got sub-atomic electron orbitals in a systematic whirl of energy, mass, motion, force fandango. Sure your proton spin might be clearly distinct as per a strong force spherical formation. But it's superluminal combination of spacetime and motion is probably having a synchronous effect within the system. I mean it's up for constant dispersal via concurrent spiral interaction. It's driven by the system.

So the micro force formation might not consider mass and energy the way a macro version does. It's not singularly reacting to it the way a larger version does. Again, it's the *system* that theoretically adjusts to it rather than the individual force would on this level. Remember a previous chapter about mass-motion condensation? Well, theoretically on any level a spherical spinning formation epitomises that. But on the atomic level it's not just the spherical spin of a proton that depicts a mass-motion fusion. So might the neutron, along with every other bit of sub-atomic fluff in there. (Pardon me, Plank.) Indeed the protean nature of the neutron seems to reflect the atomic rapid fire team spirit. And then you've got wave/particle dual natures happening as well. But despite all that, you can still decipher four force formations at work inside the atom. Well, I can anyway. I can draw correlations. It's still a system on the bifurcating general ladder, albeit the theoretical bottom rung. Once more, look at the distance between proton spin and electron/orbital formations. There's plausible versions of a planetary/stellar set-up there.

A macro process wouldn't express the same degree of organisation though. It's not subject to the same amount of progressive layers of symmetry for a start. As we go up the ladder we wouldn't find the same nuances of mass and energy within the force formations. Your wave is clearly distinguishable from your particle, for instance. Energy and mass don't fuse and fudge according to our methods of detection so much. We're not getting superluminal interference like we possibly get within the atom. We're not groping for neat energy and mass ratios on a level where force formations vie for their specific spacetime and motion combinations all at once. Further up the ladder, one force can more freely react to the bound atomic packages. It can then register an energy and mass ratio towards its bifurcating level, which also accords to the symmetry of a still greater force formation and the general picture.

In any case, these levels of force formations show how matter and energy might transform progressively down the systematic line. Or up the ladder if you like. And here's something. I think the reason classical physics doesn't reason through levels of force formations is because the spacetime and motion ratios remain invariable. So they could be missing the symmetrical levels, and the bifurcating nuances of the force. Lucky physicists. Gravity, for instance, forever remains below speed c with maximum spacetime on any level. According to this theory however, micro gravity—that same ratio of below speed c spacetime and motion, interacts practically simultaneously with luminal and superluminal combinations. It's thus diffusing a definable distinction between mass and energy ratios. At least according to our sub-luminal forms of measurement, that is. But get this. The more symmetry and organisation, the more intricate are the patterns. The tighter the system—greater are the density, form and amplitude of mass and energy. Your rapid fire micro fandangos come up with perfected elaborations not seen on the bigger levels. It's just too bad we can't infiltrate the fine tuning to get a glimpse of it. Here is our bifurcating ladder echoing through manifold symmetry. And that's me wandering through the thesaurus. I'm trying to analogise a glimmering current in Versailles' Hall of Mirrors.

The point is, classically, theoretically, or any way you care to consider them, we've got physical correlations reflecting staggered systems. Theoretically they represent the order of symmetry. And I think we're looking at the bifurcation of general scale forces through them. Feasibly, it happens out of your cyclical junction. Perhaps the bifurcating ladder unfurls out of a strong force formation—say a big bang. Stuff ostensibly stabilises. It cools down, and systems forthwith form. Galaxies arise. Within them, come stellar systems. You can gauge them by the similar set of force formations. Look, I'm not going into the nebulous ins and outs of cosmic technicalities with this either. Who's wearing Hubble lenses? Jesus, any technical presumption here is probably worse than theoretical persuasion. All I'm saying is these systems show the same formations of forces at various levels. You've got waves, spirals, orbitals and spherical spin on at least 4 levels. The general forces' symmetry apparently transmutes and remains invariable all the way down the bifurcating line. Rather it's the *physical* manifestation—the energy and mass ratios of these formations, and their systematic rate of interaction that varies per level. You see; as a system they directly and indirectly carry a different burden of symmetry.

I mean it's no big revelation to say the evolution of atoms concerns the evolution of stars. Each process bears down on another. I'm just proposing that these processes stagger the cyclical set-up. You can have an atomic version of gravity and a galactic version of the weak nuclear force. They're all symmetrically relative. The stellar system has similarities to the atomic one. The galactic one should correlate with the stellar one. And the cyclical one should instigate and umbrella the lot.

In short, the four force formations are subject to their own symmetrical division, which organizes on different levels. Regardless of scale though, the force functions according to a set combination of spacetime and motion. The system thus describes physical correlations between the levels. The critical density to make a galactic black hole, for instance, could approximate the nuclear density of an atom. Yet these two situations of spherical spin would maintain symmetrical balance at a different rate. Its system—the four force interaction, is subject to a scale of symmetry. Each level is more or less subject to the other rungs on the bifurcating ladder, as well as its general symmetry.

So my little feral domestic, there you have it. No matter how much it hurts, I can't change you back. You're becoming more feral than domestic, aren't you. There you go, breaking my heart by venturing farther and farther afield from its loving embrace. Sooty, the Braveheart. Come walk with me, if you will, up through the wild fox gloves, deeper and deeper into the forest each time. Watch me go up that big blackwood. But don't wait around for me when I go into a thicket of bracken. And don't cry out my name. I've got places to go and smells to smell. O.K. then, I'm getting the message my honeys. Still, I can see you sometimes. There you go, leaping and bounding up through the bracken after the little baby birdies. They don't stand a chance any more, do they. Well, my clever hunter, the eagles always come around when you're not watching for them. And it will always get cold and windy any day of the year. So don't lose your way home. Don't forget the cozy warmth of its fire. Come in out of the rain, and let me cuddle you still. And get out of them wombat holes, flea bag.

. *summary*

I suppose we should have a summary thus far, before blasting off into the superluminal conclusion. Hopefully however, we've arrived at a position to do so. By now we should be soundly in the infinite and eternal context, ready for full throttle brainstorming. On the likely off chance that we're not though, I'll detail some of the finer points.

To overcome the luminal barrier I'm reaching into the eternal and infinite dimension. One apparently presupposes the other. (Not the barrier—overcoming it.) The notion also presumes an order to the cosmos. You get ongoing transformation processes with the logic. So everything doesn't commence with a big bang. Nor does it stop at speed c. Theoretically it all comes in cycles—an infinite array of them. You get a causal process out of them. And as a process, they can over-ride a singular primal cause scenario. Plus they can cover the entire cosmos. The expansion of one cycle, for example, would be subject to the contraction of another. There's no endless expansion of bigger and bigger conglomerations of a one off system. Here we're looking at the continuous structure of phenomena—skeletal, auxiliary and repetitiously.

Underpinning the chapter is a theory of symmetry. Its logic works in all directions and dimensions. We've symmetrically got subsidiary and primal causes. *Primarily*, the cycles describe a macro dimension. This is your general differentiation of spacetime and motion. You can find a pattern here, showing four formations.

And it appears to be the basic symmetrical division of spacetime and motion. I'm wondering that it applies throughout an infinite and eternal cosmos. For the sake of balance you couldn't have any particular beginning or ending. So the primal cause—if you must have it, *is* the process of infinity and eternity. It's the reality of spacetime and motion. You don't get it without a balanced process. I'm kissing chaos and the inexplicable good-bye, good-bye with all of this. Although admittedly, you've got buckleys gleaning solid proof of every cosmic event, using it. All we're getting are some theoretical suggestions and the gall to go for it.

There's not a lot in the chapter about negating a primal cause. Not directly, that is. I'm not arguing against stuff, even if it is academic. Besides, there's argue and there's *argue*. I mean did Plato argue like we *argue* his pointers with Socrates? I thought he had a dialectic to produce a higher level of truth—a synthesis of thought. I can't see that's the same thing going on with his political descendants. Look at them in the house of reps here, assiduously and insultingly at each other's throats day in day out. What synthesis of thought? Frigging married to each other. It's enough to put you off politics. Yes, well . . . debate, argue, or yabber, who cares. I'm clearly not summarising, but splitting hairs, mooting and digressing. The thing is; we're going where our powers of observation don't take us, with this. Cyclical links mean that the beginning is also the ending of something else. So we require more than one cycle to complete a balancing act. It's that simple.

How big is the cyclical level? Our basic order of cycles is definable by the basic system of *force formations*. This is where their symmetrical completion, or a balanced set, requires another adjoining one. But they bifurcate. The orbital, spiral, wave and spherical spinning curvatures of spacetime and motion, bifurcate internally. It provides the scope for manifold change within each cyclical scale force formation. And it also creates order in every dimension and direction, because the bifurcation derives through the greater balancing act. How intricate then, can we go within a cycle? I am gauging it from the inter-galactic scale through to the atomic one.

Cyclical balance controls any rogue 'infinities' of subsequent bifurcation. And I'm talking staggered levels of force formations bifurcating out of it. The energy and mass ratios do it. With constant change energy and mass always transform according to these forces. A condensation of mass can imbalance the system and bump it up onto another level, via a greater forceful interaction. And these staggered levels check the symmetry within the cyclical system. Each level restrains the others within it. How the system functions and its consequent energy and mass transformations remain subject to it. The more intricate is the system the more the forces fire away together. And at some micro level within a cycle, force formations bounce back according to the invariance of macro symmetry. You see; the macro level is changing as well. At some macro stage these formations are subject to the adjoining patterns of bifurcation—of the cycle next door. In short, we've got a skeletal four force pattern generally, and a staggered multiplication of them subsequently the whole of which is constantly changing through spacetime and motion.

Dark matter figures in all of this too. On the general scale, it's the available spacetime and motion according to a neighbouring cycle. Microcosmically, its shortage can render force formations and energy and mass transformations barely distinguishable. In either sense, a paucity of it confuses the causal process. Mind you; I'm not suggesting the stuff is a virgin slab of non-differentiated spacetime and motion though. It's ostensibly the recycled business of previous and co-existing cycles. And you may ask; what combination of spacetime and motion is it then? There's theoretically no such thing as space without time or motion amounting to some force formation on some level. It all comes in some forceful form or another. But I don't really get into that. I just infer that it's the stuff pertaining to cyclical and bifurcating contingencies. nb 5

Bifurcation? Again, the four forces replicate inwardly. They amount to levels of symmetry, or systems within cycles. These force formations proliferate according to a general level. It could be something like fractal geometry. I don't know. What's fractal geometry? Or it could resemble a pyramid. Whatever. I'm referring to the invariance symmetry necessarily realises. Within cycles we seem to get decisive *levels* of transformation. But they remain relevant to the entire cycle. They're always subject to the changing general symmetrical effect. Which means there could be a limit to how many levels of force formations can proliferate.

And I refer to all of that as bifurcation. Symmetry happens within symmetry. And the general pattern blankets everything within it. Maybe that's not the right word. It carries it like a tree trunk. That is; your primordial four force formations individually predominate and prevail over internal manifestation. Your leaf does bugger all once it's fallen off the tree and its trunk. Even so, these internal systems can still get around their generally pervasive effects. I mean each level comprises of four force formations, albeit diminished versions. And they still fire away according to the same combination of spacetime and motion. That's an important point. Your micro version of gravity still functions according to the same spacetime and motion combination as the general one. It's a symmetrical imperative. Gravitational attraction is always below speed c. But then so do the bifurcating versions of the other forces remain true to their combination of spacetime and motion. A wave is always at speed c. A spherical spin, big or little, is always superluminal.

These levels *systematically* react to the generally pervasive effects. It can stimulate a systematic shield. Thus your atom per se might not travel beyond the spacetime and motion combination of the gravitational force. But inside the atomic shell a micro version of a nuclear spin could be firing away superluminally all the jolly time. Symmetrically, the effects of the macro level recede as we get into each new level. These internal systems tune in to the bifurcating symmetry. Your leaf answers directly to its branch and indirectly to the trunk. That means the general relativity of gravity—whilst still overwhelming—might be less decisive within atomic systems. Not like it might be with a stellar or galactic system.

And so what. We still can't distinguish superluminal stuff within the presence of the macro gravitational force formation. You've got to step inside the bifurcating level to do that. And with us, it's not apparently possible. We can't go inside a solar spin, or inside an atom without greater gravity's guidance. It obscures our abilities to see the logic of a quasar. We can't see its timely relevance to a cyclical process. Nor can we pick up the superluminal velocities within the cyclical sub-systems. We're groping around inside the atom with a bulky beam of light chewing up the intricate spacetime and motion network and freezing the forces into position via their momentum.

Then I conclude with correlations. I'm trying to 'prove' the bifurcation of symmetry with them. That's basically because correlations describe contrasts. The idea is to find differences and similarities between the levels of force formations by using them. What comes up are the energy and mass factors, and varying rates of force interaction according to the bifurcating level. Does the production of superluminal activity on one level differ from another, then? Apparently so. The micro is subject to more levels and thus greater organisation and control for starters. Your sub-atomic spin tunes into its own system more than the macro version. So we're getting there. We're theoretically on the road to inter-galactic possibilities. Move over SETI search. Mohammed's not coming back to any mountain. It looks like we've got to get the mountain to him now. And to date, it seems we only blow them up. Maybe they don't want to know us until we can figure power out otherwise.

. endnotes

1 nb ref: p.198 Atlas of the Universe

2 nb And look, search for it if you must. You could just find that it's some enigmatic outpost with even more enigmatic referencing attached to it. At some stage I probably stumbled across it in Philip's Atlas of the Universe. Be blowed if I can find it again.

3 nb Active spiral seems to evolve out of spin in the atom to figure inherently within the system. Yet here it seems the other way around.

4 nb p.200 The Atlas of the Universe

5 nb It's conceivably the available spacetime and motion according to the greater scale, staggered through to the bottom level. It's the potential for greater gravitational expansion according to the neighbouring cycle, which is also up for atomic transformation. It's recycled, and you can't recycle without it.

6 nb ref.lost ie in the diagram

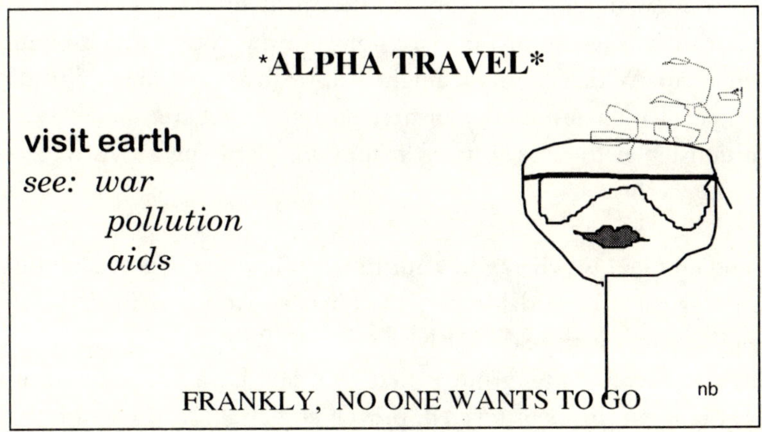

ch.9

superluminal travel

. preamble

Before proceeding, I want to come down from any astral plane I might've wafted up to through conceptual extravagance, and face facts. I want to grovel in the gravel of reality. I mean I've got to account for any loop holes in the following. We're getting into nuclear physics and all manner of wizardry next, the technicalities of which are well beyond my credentials. Look, this book is just a big theoretical step into a future world. I'm sticking with the facts as much as I can, but hell I clearly don't know them all. Who does, anyway? For all you know I could be right. For all I know there's volumes of stuff not considered. Plus we're dealing with a mathless deficiency. And I don't have any recognised skills to write all of this stuff, you know. There's no previous publications. Still, some Cosmic Wonder might capture the spirit of discovery. You can bounce off suggestions. Bounce right out the door. Or just bounce off them. Which means it might help to have a cosmological dictionary handy and a crate of beer on stand by. Sure, there's a definition list at the end. But it's mainly for clarifying a twisted sense of key terms.

The problem is; when you want to reach into the next realm you're up against a formidable torrent. First, you're up against the notion. What's with realms, I hear you ask. Bubbles, realms, we're in it, and there's nothing but it that we're in. Then you're up against not one, but loads of revered, proven and established laws of physics. Plus all the influential eminencies, and hard evidence that go with them. And it's difficult to ignore their side-effects—what society does with the knowledge. Einstein must've been in some dilemma when he knew what he knew with the maths up his sleeve. Heaven forbid—at least you're only getting an unsupervised theory out of me. So then you're up against yourself. Try ignoring your conscience. It's hard to forget these informal failings. Although at times, you salvage your pride and think—way to go. I mean—perhaps Einstein could have worked it all out if he wasn't so swamped in numbers and the realities of the day. And then you vacillate again. Theory, right or wrong, still isn't worth a pinch of anything without the skills to carry it forward. Yet neither is a brilliant set of numbers. Sorry, reader, but that's preamble for you. I feel like some sportsman, hedging around the arena, ducking and weaving and psyching up before diving into the fray. Thank Christ we're almost there. And roll on tomorrow. Won't it be pleasant to taste test another batch in the walled garden.

Anyway, despite all the doubt and foibles one thing remains clear. Cosmologists, nuclear physicists—none of them have come up with a way to exceed light speed. So you think that speaks for itself, and it's proof that no such thing exists? Maybe. But none of them have aligned general relativity with quantum theory either. Nor do I see any pre-big bang logic. I reckon the closest thing to it was when the pope suggested we shouldn't delve into that. It's an act of creation according to God. Which makes me wonder that God wants us to remain inside the bubble until we improve our manners.

I studied wars and warfare at university. It's also known as international politics. It draws you into a vortex of hard realities. You end up locked into the world news, forever clueing up on the latest conflicts. Here we go, into another glorious hum dinger that could escalate and wipe us all out. Over time, you begin wondering if such devoted attention is just upping their ratings. But you still wonder what's with the zealots. Are they going for aggrandisement, or trying for a second coming? You up there in the stars, get down here and save us. I don't think that stops it. We willingly live and die by the sword it seems. So what you learn is that it's a slow road to peace. Especially, that is, when fighting still turns over a profit.

What would we do then with an objective for *uniting* people towards a common goal? One that doesn't mean defining a common enemy. It could draw us out of our war mentality. And the less we fight, the less we fight. More chance then to check out the galactic neighborhood. Well, it's a thought. Today's reality, however, has Slobodan stomping on his people, and America bombing the bejesus out of Kosovo. In the meantime I feel like a lazy cow— sitting here in the wilds of Tasmania, with its head in the clouds, contemplating an astronomically attuned utopian society. (That's a Tassie cow for you.) Well, someone has to do it. If I don't stick my head in the clouds and ruminate on the cud, who will? Besides, the alternative is working out how to enforce international law without a law degree. And that means going back to some city, starting from scratch again, and locking into rent, fashion and platitudes. Furthermore, someones still have to construct practical international law anyway. Who's going to abide by something they don't understand in this, our postmodern, egocentric age of die-hard realists? And what state leaders come to such understanding without being involved in its construction? I bet Slobodan doesn't give a hoot. He'd evidently rather fight until he's battered into submission . . . and so it goes.

Apparently our nearest star is Alpha Centauri. Now if we travel at a million miles per hour it would take us 3,000 years to get there. And how would you be getting there only to discover a solar system as wasted as ours? We could always try Delta Pavoris, or Beta Hydri, or Zeta Tucanae instead. Well, dream on. Or get real. Clearly we wouldn't even make it to Alpha Centauri. So let's just wait here for 'them' to turn up. Besides, if they make it they've cracked the superluminal code and they can hand it over. And/or they've figured out longevity in a really big way and they can hand that over as well. This could be a win-win situation. Until someone shows up, we can just bide our time. We can rest on the laurels of general relativity and our other laws of physics whilst maintaining the rage with Saddam, Slobodan and any other serious war machine. And we can wallow in subjective pride and keep believing we're 'it' in the entire cosmos. Only we'd probably self-destruct in the meantime.

What a scenario. Here's another one though. What if the cosmos is infinite and eternal—and there's only one race of people? And they're killing each other? Even more pointless, what if the cosmos isn't infinite and eternal, and there's still only one race of people that gets the chance of being in it? And they're *us*? You must be joking. Or better still, what if there are other ones? What if there are zillions, and none of them can get beyond the speed of light? We're all here, in the cosmos, but in isolation from each other—a cosmic crowd of killers and pillagers. What's the point then? Evolution, be damned.

On the other hand—if we have any left, evolution might be a necessary consequence of an eternal and infinite cosmos. And if so, it sure looks like an elusive, selective, contingent thing—on both sides of the coin. From our part, that is, and from the cosmic point of view. Evolution could transcend us. But check out the fish. Check out the jungles. Check out the planet. We're evidently eluding it. Or are we? Apparently we can also use it, forego it, misrepresent it, and forget it. But it's there. It's in the bible. The thing is, do we focus on the wars to end all wars, and strum up another one for fatalistic historical validity? Or, do we grasp the commandments and aim for evolution. Do we get to go to the galaxies, to meet the phenomena and whiz around the stars? It's all very well orbiting around the planet. Evolution apparently offers a choice. For us though, it's probably a capitalist quandary rather than a moral dilemma. What part of nature is economically sustainable?

Who knows? Maybe there is after all, other planets of pre-superluminal killers out there. But equally likely some could evolve out of it and some could self-destruct. Now that, I find easier to stomach than the 'only us in the cosmos' scenario.

I don't think realising superluminal travel just requires some genius doing some fantastic calculations. The deeper I get into this stuff the more apparent it is just what we're up against—a sea of preamble for one. And then there's our finite thinking patterns. Your all time big bang, be it right or wrong, is probably just the tip of the iceberg.

By the way, Russia's steaming towards the Balkans now to shore up Slobodan as I write this. (And as I edit it, they're repeating everything with the Chechnyans.) Nato is running out of missiles. And there's talk of nuclear arsenal. I can feel the ratings going up. And there goes the latest flock of correspondents en route to hone in on the gore. Meanwhile, I'm gazing over the valley at the swifts circling in the clear blue air. What do you do? Rejoin a political party? Pour over the university research? Write letters? Or finish this book and trust in

a: the advancement of human nature
b: my inadequacies and ineffectiveness in that direction anyway, and
c: that the swifts won't nest in the verandah roof, **again**. Well, bugger that for a joke. They never learn. Look, I'm not sitting out there with swallows buzzing at me in their droppings. That does it.

So where were we? Oh yes, we're preambling towards some serious writing. In researching data for this book, only twice did I come across anything that even remotely recognised the possibility of superluminal speed. I think Ian Crawford, a British Astronomer, considers that special relativity doesn't explicitly forbid it. And that the laws of physics may permit it. (Reference lost, I read it somewhere.) They just don't explain how it happens. And then Brian Martin, an Australian astrophysicist, recommended I look at tachyons. Mind you; he didn't say he agreed with them. So I looked at them. But they rely on a reversal of the causal process. I mean you get the jolly effect before the cause with them. The point however, is that superluminal speeds are improbable, if not down right impossible going by the general eminent trend and the facts before us. Still, I'm curious about the special relativity case. So let me re-state something.

Theory is showing these speed barriers relate to ongoing cosmic processes. You've got to presume something beyond the finite state to overcome them. And that's well and good. But it's still largely presumptuous. You can't prove notions of infinity and eternity through the evidence before us. You never can. What evidence would it take? Yet according to this theory it seems to be the case. It's our catch 22, and the point is our laws of physics could well work in with it by recognising a contextual relevance. Remember, we did this with paradigms? Now I'm wondering that here is our special relativity loop hole. This law could predicate a broader context embracing the faster, the deeper, and the beyond. The superluminal situation should point to a highly curved space that transpires out of any event horizon. And using this theory, we can do that. We go right into the essence of curvature with ideas of spherical spin and mass-motion transfer. Resting on the premise of infinity, the black hole spin transpires into something. Its highly curved space calculates into another nonlinear form that also manages superluminal velocities—say of a quasar set-up. We're not denying the laws of physics. All we're doing is judging the next phase of events by using the bigger context. So do you want to come, or what? I'm heading for the hills on the other side of the valley—up over there where the mists are rising out of the mountains to meet the clouds. We can explore the uncharted territory beyond the finite boundaries of physics. I'm going into the nonlinearity beyond those limitations.

To get there, I'm traveling over ground already covered. But this time, let's hope, with more insight. I'm going into rotational dynamics again, and nuclear power. Complex it might seem, but really it's not. Remember, the grovel. Who's the wizard? Everything is remaining under theoretical cover. So don't expect any deep technical stuff. Nevertheless, it's probably tempting incredulity to leap over too many proven facts. I'm therefore venturing inside the techno world and putting my head in the lion's mouth. But please bear in mind, in alluding to all this stuff I'm really just skirting the conceptual guidelines with it. On that basis, we can possibly grapple with a few nuts and bolts without burning fingers. The mechanics of superluminal speed, for instance, concern some greater nonlinear situation beyond a gravitational spacetime curvature, rather than the ins and outs of aerodynamics, torque and algebraic expression. Oh, and there might be further reference to cosmic rays and accretion discs later. But if they look like rehashed pseudo bumption I'll edit them out. Great isn't it? I'll edit them out. This is the edit talk-back here, brought to you through the powers of unpublished bravado. Clearly we've come through the grovel unscathed. So let's do it—except for this.

There's finally something on colour changes and speed. Not that I understand it. It's more of a curious addendum. I saw something change both position and colour in rapid sequence in the evening sky here a couple of years ago. And strangely—or timely I suppose, I've never seen it since. Except the other night, summer time this time, a flotilla of synchronized satellites glided past, disappearing together before reaching the horizon. Synchronized satellites? There they were, three in a triangle. Makes me wonder how many satellites we're going to shove up there before we get a traffic jam.

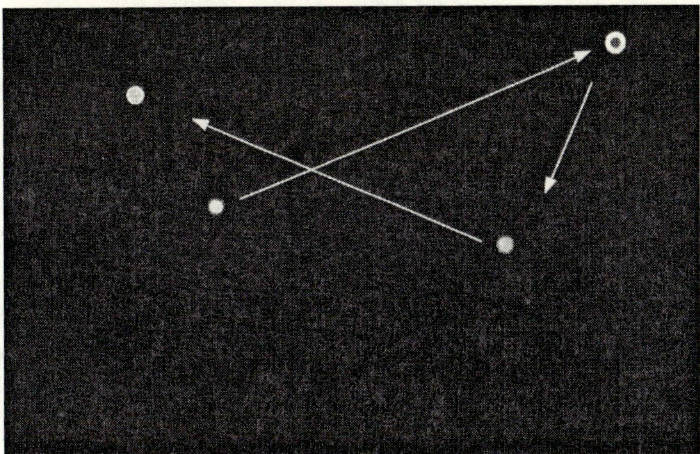

Don't quote me on the sequence.

. rotational dynamics
curving motion for more speed—forming the force

With rotational dynamics we should be getting into the mechanics of superluminal power. And I'm more than likely in way over my head. This is the nuts and bolts stuff of physics. We're coming out of the clouds to face some hard evidence of Newton and Einstein. And look, I don't really want to. SOOTY my honeys—come out of the bracken and let's shoot the breeze. Besides, the term 'rotational dynamics' doesn't come up in my cosmological dictionaries, anyway. It does, however, appear in those physics books up there on the shelf with them. Yes, there they are, those dusty old ones under that new pile of glossies on the cosmos. And on consultation, there it is: whole jolly chapters on rotational dynamics in all of them. Whole chapters that I don't feel like re-reading.

You'd reckon nuclear physics would be a singe by now. I mean this is four years into the text. In reality though, all you're getting is a dusty sniffle. What rotational dynamics mean according to those books, is stuff I'm familiar with—sure, cognizant of—well . . . not really. Actually, no. So I guess we're pushing it. Even so, there are aspects of these laws that look incredibly appropriate to theory. More specifically, they involve our nonlinearity. Your kinetic energy and rotational inertia seem to concern the inherent motion, *shape* and distribution of mass, for example. And motion is a relative thing. It's looking challenging then. But can we really link an abysmal comprehension of the data to superluminal possibilities that have, as yet, not a shred of evidence to their credit? I mean they don't according to scientific experiment and observation. You don't think I'm living up here in the hills conducting strange scientific experiments with these skills do you? Home brewing and gardening—that's about it. So read on at your peril, and become either enlightened or confused.

According to theory thus far, rotational dynamics would concern the curvature of motion and how that can manifest power. I'm judging the curvature of spacetime as motion. You don't get spacetime without it. So force here is a *formation* of spacetime and motion. A ratio of motion combined with a ratio of space and time expresses the formation. There's four of them. As the symmetrical division of spacetime and motion they form a device that then figures according to mass and acceleration and thus the rotational dynamics of classical science. So there's rotational dynamics, and then there's evidently rotational dynamics. In other words, normally we only judge rotational dynamics according to mass in motion. Force, as I think we generally understand it, is a matter of Newton's laws of motion, namely $F=ma$. But is that equation just reasoning consequent motion and matter density according to greater gravity's general presence? I mean this could be where the inherent motion of spacetime curvature hardly figures. It could be where spacetime is barely a nonlinear event to start with.

I'm talking the nonlinear situation of each force, regardless of the mass/energy transformation. And I can just hear it. How can you distinguish a force from a mass/energy transformation? Right, well maybe you physically can't. But at this stage of the game I am. I'm judging it by the invariable nature of its symmetry. Remember, these forces ostensibly remain subject to the symmetry and bifurcation of systems and greater processes. And for this reason I don't think you can gauge them unilaterally according to mass/energy transformations. If we do that, we're probably back to general relativity and nuclear force non-alignment. The only way I can reason these forces on the one scale is according to the invariance of their symmetry, which is their unchanging combination of motion, space and time. It theoretically never changes despite their systematic bifurcation and energy and mass transfers. This is where motion is relative to spacetime at its most fundamental level. We're not just talking about greater gravity. Here we arrive at the force through a curvature of motion according to a particular combination of space and time. Then you figure it out with mass/energy according to its system.

Like Einstein, Newton's probably gauging stuff according to the general relativity of gravity. That is, according to the invariance of the prevailing cyclical-scale force, which happens to be gravity. I don't think you could apply his equation to a quasar process, for instance. Indeed, unless we predicated it with another equation, you mightn't get around to reasoning a superluminal situation of nonlinear mass-motion fusion on any scale, cyclical, atomic or otherwise. His acceleration comes over as a linear event to me. And I think he arrives at it through the length and spacetime of a macro gravitational orbital formation where nonlinearity is at a minimum. Everything goes in a straight line until acted upon by a force. You're not getting the motion of spacetime curvature. Indeed you don't get motion at all until mass comes into the picture and acts according to a force. So is there a way to apply this equation to the inherent nonlinear dimension? Can we still gauge the nonlinear dynamics of the force despite the linear imposition of greater gravity? Can we translate Newton's logic into a more profound context? Can we apply a metaphysical rationale of the force with the proven logic of energy and mass transformation

Without going into all the bifurcating ramifications I'm giving it a go. Look, even without the maths be buggered if I'm going to try and figure out all the rotational nuances of galactic/stellar/atomic or cyclical facets of the force. Besides, we've already got the cyclical facet of the gravitational force anyway, supposedly with F=ma. I'm keeping it simple. And so it is. First we've outlined a metaphysical grid for four forces of spacetime and motion. To give that physical meaning I'm bringing in Einstein's energy/mass equation. Next I'm translating Newton's equation according to the grid and Einstein's equation. On that basis we're putting F=ma into a non-linear dimension that encompasses and also transcends gravity. It's where motion can figure more prominently within spacetime.

Here we go then. First, let's re-arrange the definitions. I mean we're talking the primordial level of the melting pot at this stage. What happens if we allow mass to translate into motion on the spacetime side of F=ma? I mean if spacetime takes up the **accelera**-tion side of things, maybe **m**ass can take up the motion side of things. Force equals mass/motion by acceleration/spacetime. If how-ever, we're going to do this, if we're recognising a mass/motion con-version then we're also recognising an energy/mass transformation. Mass/energy would be just as convertible as spacetime and energy.

Next, using the proven $E=mc2$ equation we're covering all the transformative possibilities of the four forces, despite the bifurcat-ing symmetry. That's giving us all the physical dimensions of the four forces. Using it, allows us to translate stuff, most notably the time of Newton's acceleration up to speed c squared. But I'm not just talking time here. I'm re-arranging the $E=mc2$ equation to the breadth and depth of the spacetime and motion metaphysical grid, not just according to the minimal motion and maximum spacetime of greater gravity. And it seems to work out. Look for instance, at the mass increases we expect at high velocities. Theory suggests the strong nuclear force has a maximum quota of motion with a minimum one of spacetime. Accordingly, when a concen-tration of mass figures with a minimal amount of spacetime 1n b—

say tiny space in time squared—it could amount to a very power-ful force: the strong nuclear force. I think we're now looking at spherical spin. And such spherical spinning mass could translate into the Energy of mc2.

Classical F=ma doesn't factor everything in this way. That could be why we only measure the up spin or down spin of a particle. The particle would probably involve either spiral or spherical spin-ning motion. And the up and down, and every other relevant part of its spin, should tally up to something beyond speed c. Yet that would require consideration of a mass/energy conversion within the non-gravitational flexibility of E=mc2. Our up and down mea-surement could be the segments we're getting by way of the linear rationale of general relativity. That is; we're only judging the force according to the invariant possibilities of greater gravity.

The thing is, applying rotational dynamics like this—right from the onset, might seem to be undermining the linear rationale. It might seem like we're undermining the general gravitational fac-tors of mass and energy. But I think all we're doing is putting the linear rationale into the greater nonlinear dimension. We're not just gauging the force according to gravitational tenacity—accord-ing to the predominating stage of gravity in the cycle, but accord-ing to the abilities of the force per se. Be it inside the atom, the galaxy or the solar system. I think to recognise the symmetrical prevalence of greater gravity we just follow suit with the F=ma equation. More or less. I mean force equals all of this business above, which according to its system, then factors by way of mass times acceleration.

I suppose what I'm suggesting is the rotational dynamics of the force and the energy/mass transformation is not just a question of greater gravity. When you look inside the atomic system you can theoretically figure out these things regardless of greater gravity. Theoretically greater gravity only affects the atomic system per se. The shell generally gravitates rather than the internal wizardry.

But take away the invariant nature of greater gravity's symmetry, or the imposition of the linear dimension, and motion is as much a part of energy/mass conversion as space and time. Indeed, it can factor over and above spacetime to the tune of speed c2. On this basis, matter density should tie in with the nonlinear intensity of spacetime *and motion*. It doesn't though. Going by our laws of physics, when matter density is too extreme we get a black hole situation overwhelming gravity. And that's the end of the equations.

We don't gauge variations of mc2 into E and then E back into variations of mc2. We don't go; spin, spiral, wave, orbital on systematic levels. We can't gauge stuff except through spacetime curvature of greater gravity, be it on the general level or within the bifurcating sub-systems. And the formations we see, on the atomic, stellar and galactic levels don't ring any bells. Compression at the black hole stage should theoretically mean the mass density is generating a more highly nonlinear force. Spacetime should be curving into a tight knit spherical spin where mass starts factoring as a fusion of motion in spacetime. We could be entering the rapid realm of E=mc2 in a spacetime fizzer. This could be full on strong nuclear force formation stuff on a prevailing scale.

You can arrive at the same logic inside the atom and those other systems. Matter evidently transforms into different forms of energy on different levels. These variations can attune more to a different force formation than gravity, give or take gravity's greater presence according to the scale. Take uranium. Does it involve the weak nuclear force more so than the gravitational force? Does uranium release powerful energy because it comprises of a nonlinear condensation of spacetime *and motion*? Could this amount to a tight curvature of motion whose power we might attribute to the spiral force formation?

Again, science tells us mass increases at high velocities. But it can also become energy according to E=mc2. So it could be something incredibly nonlinear. It could invoke a force *formation*. Take gravity per se. We're at the other end of the force spectrum with it, where *mass-motion* is minimal to the *acceleration or spacetime* part of the equation. Indeed, it's so minimal that you can apply F=ma. It looks like motion can translate simply as mass and requires a lot of space and time to amount to gravity. There's your orbital formation—the most 'linear' out of the four. And try the wave format. Equal amounts of mass-motion and spacetime multiplied by each other to realise the zero rest mass and energy of the wave force formation.

I'll try to explain it another way. On a linear framework we clearly know the speed limit. At it you either stop, or as some would say, go backwards. Or you end up in an inexplicable situation. And this is the situation I'm getting at. Or, if you're not with me, already in. This is when we have something more nonlinear than a wave and an orbital curvature of spacetime and motion. This is when you're not going in a straight line until something acts upon you. We're not just talking greater gravity here. And we're not talking home brew. Theoretically, if you can't transpose space and time with motion, if you can't figure it out beyond the rotational dynamics of an orbital or a wave, if you can't figure it out beyond greater gravity's imposition, then you won't realise any speed beyond c. We don't really get into any nonlinear situations with the wave or the orbital. You can't go through, under, over and within, faster and faster into more intense nonlinearity until motion supersedes the space and time of the situation. That's when motion seems to be so intense within the diminishing value of space and time it calculates as mass. This is when we're not just talking a zero rest mass situation. This is when we're at our theoretical fusion of mass with motion.

They need nonlinear detectors for cosmic ray showers, you know. And these things apparently are the fastest measurable things going. But any nonlinear component of the detector probably only gets as far as the wave curvature of speed c and the rotational dynamics of F=ma per se.

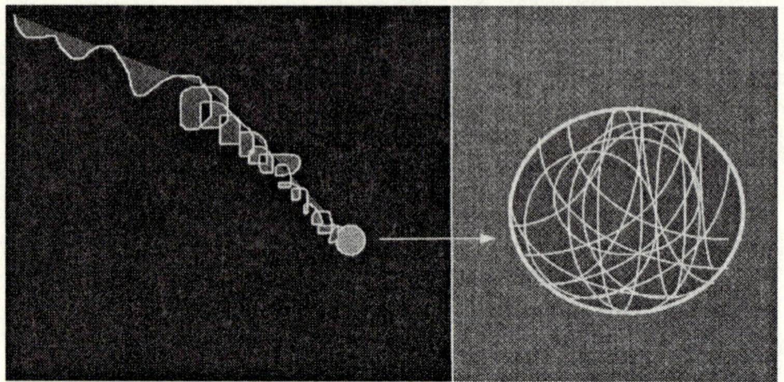

Remember, where mass and motion translate more readily it sig-nifies greater nonlinearity? So it goes here. The higher quota of mass-motion in any equation should signify more nonlinearity. You're getting less length in the acceleration bit of Newton's equa-tion. You're going beyond gravity's reach. Theoretically, that's when we're heading towards the tightly knit spherical spinning formation.

The superluminal stuff I'm talking about concerns spirals and spherical spin. They're primarily nonlinear. Theoretically, they comprise of motion beyond speed c and less space and time than the wave and the orbital. Theory thus far is showing that the more nonlinear is a situation the more motion it has. Using a nonlinear form of power then, or force, could propel things faster. Even so, we're probably only going to propel it as far as greater gravity allows, if we don't factor the basic $F=ma$ into this bigger rotational dynamics equation. So I guess the idea is to contextualise rota-tional dynamics. In short, I'm gauging the dynamics of the force per se according to $E=mc2$. Then I'm accounting for $F=ma$ and the scale and system of the force. I think it's on that basis we might be able to get around greater gravity in respect of greater gravity.

The circle below on the right shows more nonlinearity than what's on its left. The blue lines depict the fusion of mass and motion. The grey areas are the space and time.

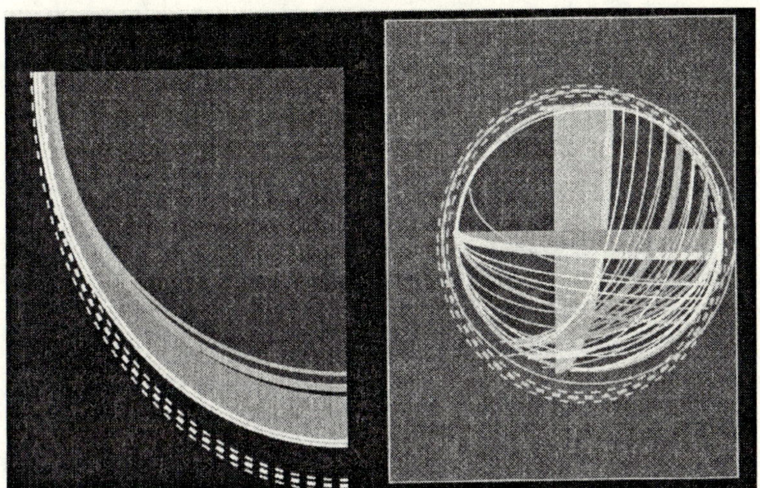

So where's the neat table of these hypo equations? Look I'm not doing them. I'm just setting out the conceptual guidelines for them.

. *the wave and the orbital*

How do you curve motion then? That's what it seems to take to realise fantastic speed. And if indeed there are rotational dynamics of superluminal proportions superseding greater gravity, is it applicable? Can we do it with what we've got? Is it even further explicable without equations? Can we hammer out something tangible with theory thus far? Well, I'm giving it a go. I'll start with the orbital and then work into waves. Perhaps we can then wend our way into the white lightning of spirals and spherical spin.

Take particle accelerators. Already, we could be manifesting power through the curvature of motion with these things, namely by way of the orbital formation and wave input. I read that you need 400,000 orbits of a synchrotron machine to raise a proton's energy to 1 million million el.volts. (Don't know where I read it.) What a waste. It's going round and round in some gigantic circle for all that energy. Imagine the voltage we'd get if the particle's motion intensified into less space and time, or length and linearity. You wouldn't have to build such a monstrosity for starters. Imagine it traveling in the wave formation. And then what about spirals and spherical spin? We could even end up replicating the internal dynamics of the jolly proton itself.

Yes well, there's nothing like the uneducated bliss of cosmic inspiration is there? I suppose all that just entails working out the E=mc2 of the process with force formations and Newton's equation, and raising the cash. Billions I believe. These things no doubt cost a bomb or three, or is thast just the initial outlay? A *spiral-spherical spinning* particle accelerator ideally would generate power plus and beyond . . . eventually. Too bad we've just done 500 million on another Martian stuff-up. Anyway look reader; at least I'm factoring in some hard evidence here. This isn't only pie in the sky creativity. This is facing reality, albeit with its hidden agenda. There's scarce chance of a spiral/spherical spin particle accelerator eventuating on the basis of this book. So I can rest on my theoretical laurels and cock a snook at observation and experiment. We may never know if we never do it.

All that aside, the orbital particle accelerator apparently functions through the wave format. At least some elements of it might. I mean we could unwittingly be using the E=mc2 factors already through force formations. I read that a magnetic field controls the particle's energy increases within its orbital whiz by containing the orbital formation. Now on that basis, mass and motion might both be subject to the direction of acceleration as well as its formation. O.K. then I'm clearly no particle accelerator authority. But who's to say we're not containing the formation—the orbit, with a wave element? The magnetic field has wave inferences.

Without getting too technically presumptuous here, I understand the size of a particle's orbit in a magnetic field apparently remains proportional to the magnetic field strength. The wave is factoring in. The magnetic input could be affecting the space and time quota of the orbital. Let's say it's supporting it. According to theory, the magnets would be controlling the direction of motion by manipulating spacetime. Without our magnets—or wave input, the particles could spiral out of orbit at high speeds. Remember, the formation stands for a decisive combination of spacetime and motion. It has its own unique rotational dynamics. In other words, using magnets seems to reduce the speed by containing the direction of motion. It's controlling the degree of linearity—or nonlinearity depending on how you want to view it. Anyway, by controlling the direction of motion we're also controlling the energy and mass ratio of the particle and its propensity towards a nonlinear fusion with motion.

And here's my theoretical crux on the event. This process could occur naturally. Indeed, it should by right of theory. From woe to go I'm basing all of this on natural phenomena remember. This nuts and bolt's grapple is really only in lieu of equations, and to refer it to present scientific experimentation. So you needn't think I'm forever in some Magellanic cloud with the rationale. The thing is; the orbital particle accelerator process might hold analogies to the orbital electron. Our charged particle in its accelerator is supposedly subject to the magnetism of electromagnetism. And in both cases the wave format is controlling potential mass increases. I mean, I think we're at the cutting edge point where $F=ma$ meets the greater dimensions of $E=mc2$. We're not just talking greater gravity here, we're in the metaphysical grid with this now.

Technically the 'wavelength' of a minute particle should decrease as its momentum increases. And momentum depends on the particle's mass and velocity. If it's heavier it has shorter wavelengths. That also means its mass could be fusing into the velocity of spin. It's going into superluminal transfigurations. Yet this wavelength still has something to do with Newton's acceleration. Try translating acceleration into length and time again, and that into space and time. A shrinking wavelength could be transpiring into a higher degree of mass-motion. It's going nonlinear. Less length can mean less space and time for the mass-motion factors of the particle. Your tadpole is transpiring into the egg. Eventually the particle wave could spiral into the nonlinear fusion of mass and motion. If so, it's possibly transcending the F=ma and going into the E=mc2 context. We're stepping outside greater gravity's reach in the rationale. Its mass is now a matter of the beyond speed c forces of nonlinearity. And this is where Newton's acceleration of length and time might not add up. You see; you'd have to remain in orbit for that.

Remember the stuff on charge? Spirals of weak nuclear force formation transpire through an imbalance of nonlinear mass-motion within the proton. The spirals then translate into waves of radioactivity. And waves in turn stabilise the negative electron in its orbital. Our magnetism is providing a buffer corridor. It's this imposition that seems to control any mass increase by the orbital electron. And here we have the same two force formations interacting with each other—the wave and the orbital. Naturally, the entire set-up gives us the systematic process of atoms. Within the atomic shell, all four force formations are firing away together, controlling particular mass and energy transformations through different levels of motional intensity, right up to speed c squared. The process amounts to an atomic compound that accords to F=ma and the general relativity of gravity.

The point however is this; I don't think we can accelerate particles up to speed c and beyond simply through the orbital rings of particle accelerators and the manipulation of the wave formation through magnetism. Similarly, were we to excite atoms, and then sit on the electrons changing orbitals, we probably wouldn't go beyond speed c. Not even if you were sitting pretty with a postmodern, light weight, stream-lined bottom. I don't see how you could whilst the distance between orbitals remains subject to a wave formation. Still, the accelerator logic shows how to control potential mass increase. So we could be on the way with this orbital plus wave input.

But why stop there? Optimistically, there's always an atomic *system* right in front of us comprising of nature's own brilliance for guidance. Look it doesn't just involve orbitals and waves. We have indecipherable spherical spin and spirals as well as noticeable wave/particle duality. They're all working around the clock controlling mass and energy levels. We've theoretically got motion levels up to speed c squared within this system. And these systems replicate onto greater levels. How well they operate beyond gravity's reach seems to concern their scale and *systematic* rate of interaction. What's more, some facts seem to speak for themselves. The highest energy cosmic ray particle, ever apparently observed for example, is supposedly 100 million times more energetic than the most powerful particles generated in particle accelerators. I mean it's not simply whistling in the orbital wind. At some stage it was probably dancing to the beat of another formation.

Imagine again then, if we operated a particle accelerator as if it were a systematic process comprising of the four forces and their formations. What would we arrive at if the accelerator took on the spiral and spherical spinning formation—per se or together? The ability to blast the entire solar system to smithereens in seconds? An implosion of credulity at Cosmic Wonder? Total disbelief? Now there's some realistic optimism. I could go into Happy Hour on that note. The verandah is looking really appealing. So appealing, I could just sit there watching the sun slide towards those mountains and consider the shadows drawing over the valley, instead. Sooty's out there. Cheeky naughty is inside the cover of Ken's arm chair—he's in its purple tent catching the summer breeze and bird song.

. the spiral and the spherical spin

At this nuclear stage, things could get rather stormy. So brace yourself for some full on deep superluminal wrangling. Brace yourself for total befuddlement perhaps. The good news is we're nearly at the end of the book. The bad news is that when you get there you could be mind swept into the nonlinear suspension of superluminal misapprehension. Or hanging out for some spiritual stability. It could drive you to drink. Fortunately for me I'm there already. In fact, if anything the book is having the adverse affect. Try writing this stuff with a hang-over—you must be joking. Anyway, the only brainstorming coming up is in so far as we're at the pivotal point of theory meeting the cold faced tide of scientific facts. I'm going to toss around important technical items like flotsam and jetsam. We're now in the nuclear land of fission, fusion and synthesis, and this stuff stratifies way above the clouds surrounding my head. Proceeding therefore with only a rudimentary grasp of nuclear physics might seem arrogant in its most naive form, but it's the only way open from here on in. Sorry to keep saying it, but I can't speak with authority on these terms. They are however, unavoidable. Consequently, I'm putting them in my bottomless bucket of metaphors to present their hues through the spectral light of theory. And talk about waxing overboard lyrically. Clearly another form of procrastination is creeping in. Again however, the idea is to fit theory with science—as can be understood by yours informally under the symbiotic guise of the rotational dynamics of $F=ma$ and $E=mc2$.

Now it looks like there's two sides to the story. On the one, by manipulating four force formations, particularly the spiral and spin, we might realise velocities beyond speed c. We could create it. On the other, this energy is supposedly already there. So why manufacture it? Let's just tap into it. Nothing to it. All it means is unraveling the mass and energy of c2 to travel at c2. And I'm wondering that nuclear fission concerns the latter, whilst fusion involves welding force formations. In any event, energy inside the nucleus is millions of times greater than we presently access. And that's what we're after. Fission, fusion or nucleo-synthesis—whatever the method, the idea is to get it.

When atoms break apart they can apparently produce 'unimaginable' energy. Technically, that is. Theoretically that's no great leap of imagination. Anyway, fission could unravel the nonlinear *fusion* of mass and energy. And this is the stuff that can yield power for superluminal propulsion. But what we achieve with fission is possibly only a microcosmic release of potential power, until we understand the heart of nuclear fusion. Let's look at fusion then. Does it involve our theoretical mass-motion conversion? And could unraveling nuclear spin require little energy—yet still release loads of it because of a nonlinear mix/mash of mass-motion? I'm talking about the break up and formation of the *essence* of strong nuclear force power. I'm talking the proton spin more so than the nucleus per se. Theoretically we're addressing our nonlinear mass-motion stuff covering c2 motion. How does it form? What forces fuse it? The whole jolly lot according to the bifurcating logic. So ripping it asunder, as in fission, is probably inadequate. I think if we're going to control and take advantage of the conversion of mass into motion we first need to understand the pattern we're attempting to unfurl.

Fusion as we know it, causes greater disintegration of matter than fission—apparently. Even so, as we know it, it still only turns about 1% of available mass into energy. Now that's not enough for what I'm proposing. You're barely touching the tip of a nuclear iceberg—let alone threatening a nuclear ice age here. And despite the gargantuan dangers of misused power at stake, I want to tap into the nonlinear spherical spin in a way that doesn't simply translate some excess binding force into power. I want to unravel the whole jolly lot, or at the very least a large whack of it to get the full force of nonlinear power via motion. Then I want to find out how to keep the ball rolling. I don't think the fusion of nuclei is that process entirely. Nor do I think particle physics gets to the heart of the matter. I'm talking nuclear *formation* according to a system. This is the fusion of spacetime and *motion* into a nonlinear set-up of mass and motion. And it's based on the symmetrical invariance of four forces. The heart of it conceivably amounts to the force of c2 that a proton within the atomic system seems to embrace.

And it's not just in the proton, you know. The bifurcating logic of symmetry indicates that a strong force situation is present on at least four levels. Even technically they know nuclear fusion occurs when material is dense enough in the sun. And more specifically and importantly, this could apply as per a nucleus. I'm just suggesting that fusion is relevant to the *spherical spin* of strong force formations, and spirals are probably the number one assistant. Theoretically, spherical spin and spiral factors of force formations exist in the big and the little of nature's processes. Nuclear fission, on the other hand, seems more elusive in nature. Indeed, for all intents and purposes it could be a man made event involving bombarding a nucleus with a neutron.

The thing is, fission or fusion, either way the release of energy as far as we can tell comes from the nuclear binding energy. Now as far as I can tell there's more to it than binding energy. But because I'm trying to fit theory with the facts to date—the ones I can grasp, I'm confining the analysis to the atomic system with its binding energy for starters. Be buggered if I'm going into the sun, or a galactic quasar with a bucket of metaphors and presumption. Per nucleon then, this energy release apparently exceeds the original. Consequently and theoretically, the role of the neutron—and let's face it, the gluon too—looks relevant. Remember, this magical little number? The neutron seems to have incredible powers concerning mass and motion formation. It holds ramifications for weak force spirals, and yet presents a point of interaction for the strong nuclear force and the proton. Perhaps our neutron can weld the formation of spherical spin together by tapping into its nonlinear mass-motion situation. You see it's got mass, nearly as much as the proton. Yet once outside the spherical formation of a nucleus, it transforms into a proton, electron and anti-neutrino. Is this our spiral link to spherical spin? Could it be a malleable agent of unravel and reformation? Does it work both ways? And heading for the heart of spherical spin, would the gluon do likewise?

Unfortunately, you wouldn't know with the gluon. Particle physics is subject to the speed c factor. So gluons, muons, etc. wouldn't reveal the whole process of spherical spin as theory has it. Indeed, fathoming the process via particle differentiation for me, is akin to piecing together a jig-saw puzzle in a spin drier. All those weak force interaction particles and the anti-neutrinos of neutron conversion—and neutrinos of fission fame—could just be two sides of the same neutron coin. Bosons, fermions—theoretically, they all involve weak force spiral formations. They all function according to the superluminal reach of mass/energy nonlinearity. And this is the stuff of mass-motion fusion—nuclear fusion. This is what I'm getting at. Their spiral affiliations should relate to more spacetime and less motion than the strong force stuff. We could be looking at the subtle differentiation here between nonlinear mass and motion through the neutrino and its anti partner. We could, but we're not.

Subject to the laws of physics, these little numbers would only show up a light green, pink or yellow flash of the whole spectrum within—so to speak. You might get an up spin, or a down spin, or something called charm. You wouldn't get the symmetrical nuances and balancing imperatives of a pattern in nature that only comes to light by way of systematic mass and energy transformations over and above greater gravity and F=ma.

Furthermore, I don't think unraveling mass and motion is simply a matter of bombarding the proton. I know that a nucleus can break apart by fission into 2 smaller nuclei releasing 3 more neutrons. I read about it. Each fission of an atom produces neutrons that continue the process. It's like the neutron is the glue holding the atom together. But is it also like a spiral that can tighten up or unravel? Fission as we know it doesn't seem to be the way to go. Sure, that might churn out various forms of spacetime and motion. But I think we're only getting bits and pieces of stuff, which only express bits and pieces of the force formation. And then we go on to quantify them as incomplete expressions of spacetime and motion, anyway. And when these things cancel each other out on collision, we could be witnessing bilateral spinning factors of the same nonlinear coin. Rip them apart then shove them back at each other, and when they collide—whoomfa. Instant energy out of mass.

Theoretically, the strong force formation has a combination of spacetime and motion amounting to a minimal quota of spacetime, and the maximum one of motion at speed c squared. The whole of which forms a mass-motion fusion. It's a decisive shape. I'm reckoning it as a tightly knit sphere. Well, these quantum particles could all represent *some percentage* of that form of nonlinearity. More than likely they're some form of its spiral affiliation. That's your strong force when it's watered down with space and time—it becomes a weak association. Call it binding energy or glue, whatever you like, but the stuff we're gauging is not the whole spherical set up. We're probably looking at a secondary mass-motion fusion in the spiral crowd of particle physics. And sure, any mass/energy combination we calculate according to them would still exceed speed c through the dimensions of E=mc2 and our metaphysical grid. But it's not the full release of strong force power. I want to tease this stuff out according to the natural wonders of the atomic system. If we shock it into reactionary particles we could also be locking ourselves out of its hidden potential.

Even I can tell that scientifically we're addressing a more aggregate situation with nuclear fusion. And despite all of the spiral ins and outs, the neutron could well be an agent to instigate a differentiation between mass and motion within the strong force compounded set-up. The thing is; the neutron doesn't seem to figure within the basic spherical spin of a proton. And this is what I'm taking as the blue-print of micro scale nonlinear fusion between mass and motion. I'm figuring the proton according to the strong force spherical formation.

So you might be thinking by now that consultation with those texts on the book shelf over there, the glossies and especially the hard-backed dusty tomes that are making it over here, aren't helping. So I might be thinking that. Who's the bewildered one in need of strong theoretical comfort, anyway? You see; the charm of a quark and its coloured counterpart still strike me as being speed limited chips off the spherical block. Sure, all up they might amount to a superluminal proton super-string of mass-motion. To continue analysing them though, individually or en masse, would probably render both you and me completely nonlinear. The edited out neutron-anti-neutrino effort was debilitating enough. Nevertheless, I can't shut the door on them. These strange quantum particles still seem to connect with theory. Does their propensity to disintegrate so rapidly have something to do with their tiny quota of spacetime? Do they reflect the shocking measures of a spherical spinning clam up?

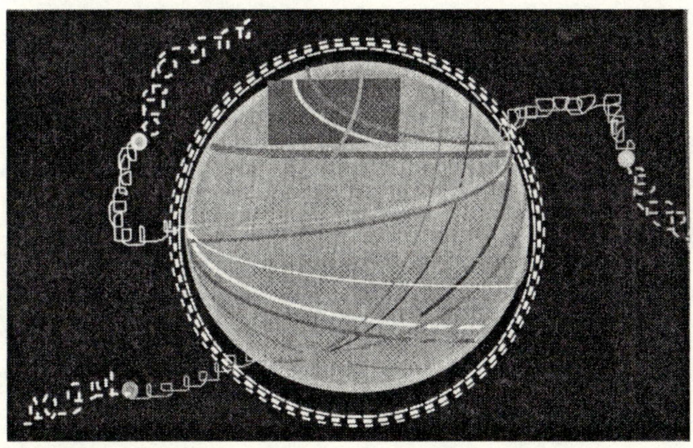

Does that make any sense?

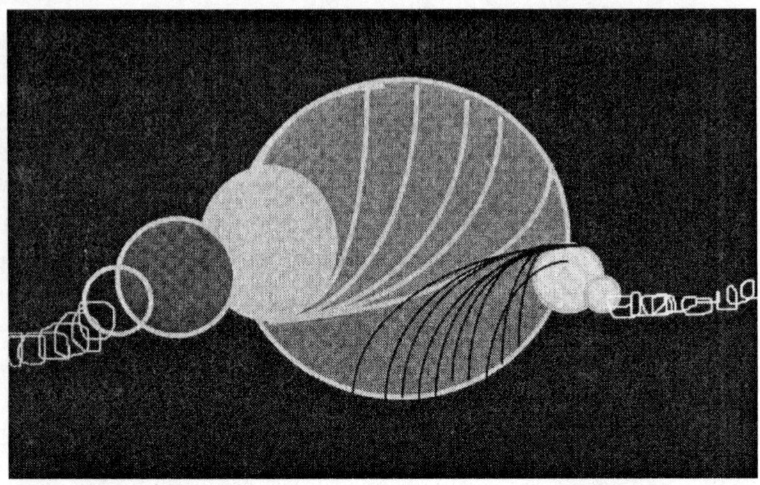

The above figment of imagination describes quarks coming out of a spherical spinning formation; say of a proton. The yellow one is the up spin, the green—the down spin, and the pink—the charm character. So how true to reality are they? Green, pink, or mud brown, I wonder they'd only ever describe transitory traces and effects of the nonlinear mass-motion through which they derive. Maybe with the quark, we're only picking up what we can get through particle bombardments, according to a speed limit. Nevertheless, quarks could still reflect discrete combinations of mass and energy according to the *force formation* that itself is subject to a spacetime and motion quota.

Either way, that still doesn't explain how we transform the mass out of an aggregate state into motion. We're certainly not unfurling the full force with jolly quarks. And any reading of them reveals nothing about a proton's release of manifold motion. I prefer the neutron. It's a natural event with theoretical backing. Its spiralised machinations seem both varied and flexible and suggest the proton spin could unravel by way of a spiral. I'm wondering that we should go in as we mean to come out. A spiralised form of energy might not shock the fusion of mass-motion nonlinearity. It might just rev it up into an imbalanced state. Enter then your neutron plus offspring with its different degrees of spiralised machinations. Consider its electron. There's more space and time in that than you'll find in a neutron's proton. You're getting greater length and less acceleration. Which means it could attune more readily with the wave side of the spiral. Whilst at the other end of the neutron range, a spherical spinning side of its spiralised possibilities may relate more to the neutron's proton.

Look, if the neutron just spits out electrons we could be spinning it up. We're drawing out the spacetime of its situation and rendering it more nonlinear. Its increasing motion could fuse into the spherical spin of mass. I mean absorbing spacetime out of the spin could even see a reverse situation of radioactivity. But if it spits out protons we're theoretically unraveling it. I guess the idea is to tease it into parting with a proton of mass content whilst manifesting the electron within. Now is that some manipulation of the nonlinear fusion of spherical spin, or what? Don't ask me how to do it though. How do you isolate a neutron then tamper with its transformative potential? Don't know. And don't ask me what sort of accelerator you'd need if indeed you did it that way. All I know is that it should surpass the big round one. But the point is, if we could get the ball rolling out of some form of neutron fission we might also realise a force formation fusion and nonlinear differentiation. That's fusion out of fission. We could end up transforming nonlinear mass-motion into mass and motion. And that's not forgetting, to do this, somewhere along the theoretical way we'd also be considering three other force formations. I know we're beyond the orbital and wave with this, but spiral and spinning sphere per se, probably wouldn't do the trick. Theoretically we need the entire four, just like the atom.

And fission—well, is this the way to go? Fission, as we know it, means particle bombardments. As a process it smacks of shock rather than the steady control of a mass-motion transfer. Besides, your bombardments probably over-ride four force formation involvement and systematic input. I'm angling for some set-up to maintain the spherical spin on one hand, whilst unraveling it on the other thus realising a continual outflow of superluminal power. Perhaps we're looking for a radioactive process alongside its nemesis.

Everyone's working on fusion, you know. Apparently no one's been able to harness fusion power; let alone the stuff I'm getting at. But they're all working on it. Nations apparently co-operate for this. So there you go. It's got something going for it from the on-set. They reckon it could amount to man's most important discovery. Stands to reason. Nothing much else seems to have united nations thus far. And who's counting on the next generation for that. We've got kids trying to blow-up high schools, gun wielding juveniles, suicidal nomads, legalised junkies and multi-sexual hedonists. Where do they get it? The family demise? Head banging music? Violence on TV? I'd even suggest there's a trickle down effect from our nuclear efforts by way of destruction. There's surely an easier way to unravel it. Some way that is, that doesn't smack of manifest destruction nor send out invidious signals to all and sundry. Beware out there; we're on the move; up and out of our bubble, like a virus.

Maybe it's that simple. For us to generate the power of nuclear spherical spin do we just put our heads together on fusion? How can you clarify motion from its nonlinear fusion into mass? Nothing to it. I'm looking straight at nature and its processes of four force formation interactions. Theoretically, I'm rising above the present reckoning methods. Well, not completely understanding them I suppose leaves one little choice. Even so, we're still going up there with the quantum zoo into these theoretical dimensions. I'm talking circles, spirals, waves and spheres here. And technically speaking, it's looking like the process should involve more than electrical forces and magnetic fields. And there's no point in trying to spell out any further finer technical details. All I know is that we're over-riding standard quantification with these play group building blocks. They theoretically inter-mingle up to speed c squared. They cover every angle. Indeed, fusion, fission, nucleo-synthesis, and any other form of high tech wizardry of particle physics would still apply. In theory, without these simple dimensions all the sophisticated stuff we're realising to date foregoes the superluminal subtleties within.

I see those particles of particle physics—the neutrons, alpha particles, etc., in terms of spacetime and motion and a force formation. If we incorporate our knowledge of particle physics—nucleo-synthesis—all that, with theory we might be able to generate the spherical spin whilst controlling its release of mass into motion. That's the ultimate objective. When I think of electrons circling a spinning sphere in their orbitals, I'm thinking the interaction of force formations. There's your wave and the orbital, affecting the spinning sphere. Accordingly, the particle—the electron, is subject to force formation interplay. Its formation derives by way of definitive combinations of spacetime and motion. The consequent interactions produce transformations. That is; they also realise mass and energy quantum. I think it's only then that we discern the electron. So fundamentally, the mass and energy of the electron are subject to an interplay of spacetime and motion forces. And indeed, working as a *system*, the four force formations all control the mass and energy percentage that basically manifests through spin.

We might need to apply the particles we quantify out of forces similarly. How does each particle reflect the force in terms of space, time and motion? Does the formation linger on through the particle? How does it translate according to $E=mc2$? Can a quark have a spiral-particle duality? And what does the spiral-quark or the orbital-electron mean in terms of mass and energy? How does that relate to motion, particularly superluminal motion? How does the nonlinear density of motion and mass become the *linear drive* of superluminal propensity? Can different particles, through force formation manipulation, straighten it out? And what about the neutron?

The big ask, and who's got the answers. So much to figure out for manifesting the motion rather than the mass when it's siesta time in the summer. And summer in Tasmania is a rare and precious commodity not to be sneezed at. Anyway, your particle bombardment probably could never do that. This ability to amass things evidently isn't getting us there. Things like nickel-56 and bismuth-209 don't seem to be getting us far off the ground. I mean the processes precluding them don't. The processes of fission and fusion, as we understand them through nucleosynthesis and particle physics, aren't powering us out of the galaxy.

How then do you control the spin, or the nonlinear aspect of those atomic mass units we realise, to produce motion rather than mass? Even I can see that the fusion process, as science has it, seems to concern particle formation. Do we understand its formation through nonlinear forces? I think you need the systematic approach for that. The immense potential within proton spin ostensibly is also a matter of orbitals, waves and spirals. What would happen, for instance, to the particle in a four force formation particle accelerator? Accelerator or not however, the point here is the nonlinear dimension of spherical spin has symmetrical and systematic ramifications that present fusion processes seem to ignore. And there's still those mysteries of neutron transformation.

So here we are, trying to create a spherical spin, tap into the spherical spin, and manifest power through a spherical spin. At the same time I want to apply fusion, fission, nucleosynthesis and particle physics—stuff I don't understand to the process, with a head cold. Furthermore, any such process would over-ride all forms of nuclear physics as anyone knows it through our laws of physics. Yes, well, it's a lovely day out there. Think I'll step outside into that—just for a minute or a hundred, into the golden sunshine of un-edited autumn.n b2 I might just contemplate the green-ness of the valley plains below and the depth of the forest beyond, from the verandah, from that big comfortable arm chair there. Then I'll sleep on it.

. controlling the process of four formation interaction

There's a lot to this. Even from a relatively simple pie in the sky perspective. I don't want to be tying metaphysical knots together. If it's not simple, it's not happening. If it's not simple, the logic's off track and beyond my mental capacities. And if it's beyond my mental capacities there's no point. The idea all along is that free from formal strictures and mathematical quagmires one can come up with a basic theory for superluminal power. And all I'm coming up with is another stream of questions. What they boil down to seems to concern a dichotomy between present methods of particle acceleration and theory's natural whoompha. How would we set a hypothetical particle in motion according to spiral and spin, using waves and orbitals? How would we gain entry into its own spherical set-up? I want to squeeze it into a very tight spin without shocking it. At the same time, I want to draw its spinning motion out through spiral elongation. And that seems to require shoving neutrons, neutrinos, alpha or electron particles, quarks—whatever it takes into it, to push an excessive imbalance of mass-motion out of it. Either that, or mess around with the neutron.

And let's say there is a way, how then do we control the release so it doesn't come out as a stream of zoo particles, but as the energy of $E=mc^2$? I want full on superluminal propulsion. And I think the way to go is in respect of hidden potential powers. That's your systematic approach. That's accessing the particle's own intrinsic spherical spinning set-up in respect of nature. Thus far, all we do is blast stuff at it. That's the attitude.

Well, there's a point. Firstly then, approach the proton with a respectful and friendly attitude. Only on that basis might its power transport us through the galaxies. And thank Christ we're nearly at the end of the book. Evidently it's getting to me. Four years I've been at this and theoretically I'm back to scratch. The motion of spin involves a nonlinear force formation involving spacetime. It manifests the strong nuclear force glue that amounts to a mass-motion fusion. And as far as I can tell, the spherical spinning particle of the proton most closely expresses this. Theory is suggesting these force formations act as a system. The proton gets around with the orbital-electrons, which get around through electromagnetic force formation, etc. In other words, we may need a set of four force formations to make energy and mass sense of the proton and any subsequent release of propulsion energy.

What happens when we remove all the electrons from two atoms, anyway? Technically, I think we're creating a sticky charge situation. Theoretically however, by taking out the electrons we're reducing the spacetime quota of the electromagnetic wave formation. And that could create a corridor for sending in your spiral. I mean we could possibly tap into the energy of spherical spin with a spiral. And theoretically, in overcoming electrical repulsion we're spiraling the atomic particles together, presumably to form a larger slab of spherical spin. Theoretically by taking out the electrons we're reducing the spacetime buffer between orbitals and spinning spheres in both atoms. Remember, 2 electrical charges exert a magnetic field, which animates space. It's the field that makes the wave. The field conveys the electromagnetic force. Without the charge situation we're going into a different combination of spacetime and motion, and becoming involved with a different force of it. One that has less time and space, more motion, and greater nonlinearity. But that's just one way of looking at it.

Taken singularly and theoretically, the situation could be a reversal of the beta decay process. We're trying to set up a spiral link from *outside* the spinning sphere. Coming from the spin at this angle outside it, we're pumping up the nonlinear fusion of mass and motion through the spiral combination of spacetime and motion. According to particle physics however, we're probably just creating a means for shoving neutrinos and anti neutrinos back into the nucleus. The trick here however, is theory recognises spirals working both ways. On the one hand, a natural chain of events has spirals coming out of an asymmetrical spin to realise radioactivity. It's a natural event. Spirals come out of spin and go into waves. On the other hand, force formations also recognise the spiral developing out of wave. Your wavelength can decrease and up its frequency into a nonlinear situation of spiral with spin. In other words, the wave might activate the inherent spiral formation. Indeed, on face value alone if nothing else, the formation of a wave could transpire out of a spiral or an orbital. But the thing is, *wave going into spiral and then spiral going into spin.*

What I'm saying then is we might be able to reverse the radioactive process. First exit the electron or manipulate the neutron, you think? Maybe. However, in essence we could be trying to reproduce what is really just *part* of an atomic process. Put these two sides of a coin together and we could be looking at a standard natural event. That's radioactive decay and its nemesis all in one. According to the bifurcating logic, on the atomic level these formations fire away together. That being the case, to maintain atomic equilibrium these things could already be happening. Moreover, as these four force formations are a symmetrical division of atomic spacetime and motion per se, firing away together provides more scope for manipulation. Up the spacetime content of one and you could be reducing the spacetime of another simultaneously. But more to the point, our wave could be contributing to spherical proton balance one spiral way and another concurrently. After all, don't atomic set-ups decay at different rates? Its spiral-wave push could be greater than its wave-spiral push in some cases. Look, with motion in the spacetime picture nothing is static. Your atomic set-up doesn't balance in limbo. Its orbitals, waves, mass and energy levels could amount to a dynamic symmetrical see-saw. I'm just splitting the atomic process rather than the atom, to reap from what we then sow.

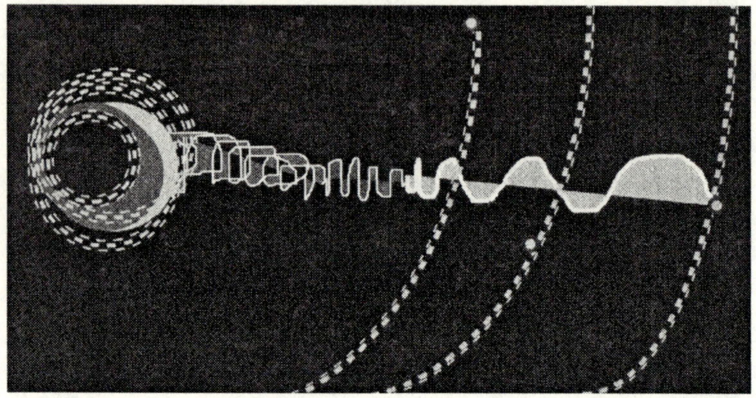

In that diagram there's light green electron orbitals happening in the dark blue and green wave area. Once we remove them, the wave shrinks in time and space into the nonlinear formation of the light blue spiral. And that's supposedly what can interact with the nonlinear fusion of spherical spin.

Now I didn't want to do this, but look every bit helps at this stage. I'm just wondering if our spiral input concerns the structure of the 'virtual particle clouds' surrounding solar protons. Here we go then, into the sun with the logic—into the lion's mouth with no technical nous. These things—virtual particle clouds—allow protons at the solar centre to go tunnelling through each other's electrical fields. It lets them do it often enough to maintain nuclear fusion. And I think that's the process by which the sun and other stars produce their energies. Heavens, the sun pours out 4 million tons of its mass every second this way. It's a natural, environmentally friendly event. It feeds the system. So there you have it. Just correlate that to the atom. Bearing in mind, of course, the structured chain of events that symmetry imposes down the bifurcating line.

Righto. It's lunch time, nearly lunch time. And I can again feel the verandah beckoning. Even un-edited I felt it beckoning at this point. The equinoctials or some other wind was happening wrecking the plants and deterring, and still it beckoned.

Our solar proton and virtual company could possibly logically correlate with some form of quark activity inside an atomic proton. If for starters, you take the spherical spin of the sun as a strong force formation correlation to the atomic one. The quark uncertainly covers bits and pieces of proton spherical spin, I think. Anyway, these infernal solar dynamics could hold some clues regarding atomic-scale strong force formation energy and mass production, and its nonlinear mass-motion fusion, quarks and all. If the solar version of spherical spin can churn out great ratios of mass according to infernal spiral activity, so might the atomic version of spherical spin, say the proton. That's one point.

We still have to take into account the bifurcating logic and external symmetrical logistics. On the stellar level, force formations don't fire away at the same rate as the atomic one. Holistically, solar energy production wouldn't directly correlate with proton energy production a la the system's waves, spirals and orbitals. Systematically speaking, it seems that the external dynamics of the atomic-sized spherical spin throw more weight around than the internal dynamics of its spin anyway. n b3 In the atomic shell, spin fires off spiral which fires off wave and orbital in splendid disregard of greater gravity. I mean the balance of one formation, say proton spin, is a simultaneous orchestration of orbital clouds, waves and spirals. All for one and one for all. Your stellar system's not showing that. We're judging the internal dynamics of spherical spin mass and energy production straight off. That is, your stellar system's spherical spin of strong force formation—ergo the sun, is having a fine time churning out mass from motion without all the noticeable ins and outs of three other stellar level force formations. Now this is not to say they don't count. They do. It's just that; at this stage, they theoretically don't obviously do that. Not as much as the atomic ones do.

Even so, the sun seems symmetrically attuned. And my point is that this symmetry is subject to four stellar scale force formations. Any internally controlled energy release should still boil down to the symmetry of the stellar *system*. The sun, for example, apparently automatically maintains its balance. As such, any *regulated* energy release through solar spherical fusion should ultimately rely on a broader sense of balance. For starters, the sun is at the centre of the solar system. You've got your planetary orbital effects. Remove all the planets and where's the sun? Is it the same symmetrical set-up? Those sun's up there without planets—well I don't know about them. But they sure wouldn't figure the same way. Planetary orbital corridors of space and time ostensibly realise stability patterns against an asymmetrical build-up of internal solar nonlinearity. Just like your atomic set-up. The difference here is the bifurcating logic. Planetary input wouldn't be quite the same as your atomic system's orbital electron input concerning the system's nuclear balance. And because it's one step up the bifurcating ladder it can encompass all the nuances of the bottom rung. Your leaf can flutter around on its branch and the branch bears fruit accordingly. Your sun can churn out masses of power via internal solar proton activity.

And there's the other end of the symmetrical scale to consider. Again, your four force formations don't fire away all at once as you go up the scale. At the stellar level, the system's internal dynamics are more open to the general relativity of greater gravity. You wouldn't find superluminal exchanges expressed as liberally as they perhaps are inside the atomic system. Indeed, any superluminal activity should remain within the solar confines of spherical spin— rather than the system's shell per se. You're not superluminal then until you are inside the sun. Nevertheless, how well these solar protons churn out the luminal result of superluminal essence is up to the system. It should still depend on the spherical balance of solar spin. And theoretically, that depends on the system's spiral, wave and orbital input. Without them, excessive build-up of the mass-motion of central spherical spin could imbalance the entire system. Even with greater gravity's general relativity *in* there, inside the system if not the sun.

Science tells us gravity pulls down overlaying layers to counteract solar core pressure. Still, despite this system's receptive nature towards greater gravity, I reckon that's your system's gravity talking rather than the general level stuff. The symmetry of the system could be counter-acting the constant solar production of protons and electrons which might otherwise overwhelm solar spin. The greater gravitational input, however, could stagger force formation activity. Perhaps spiral and wave wouldn't dance the same jig simultaneously as wave and orbital. Your solar gravitational force develops and interacts in respect of, rather than in tandem with, the spherical spin. But even if it is at a different pace, like the atomic system, spherical spinning build-up here is bouncing off release systems from within. Thus far we've got spirals, spin and gravity. For systematic validity all we need are the waves. Then we'd have 4 stellar force formations maintaining a symmetrical show of mass and energy manifestation and release.

This being the case, the internal dynamics of solar power should relate to the external dynamics of symmetry. To get our massive release of motion we may need to consider a spherical spinning situation by way of external dynamics as well as the internal dynamics of solar technicalities. I reckon maintaining a regulated release means upholding the symmetry of a system. Or at least beguiling it into an imbalance to provoke a symmetrically automated response might. The bottom line is that to control mass and energy changes we might need to maneuver the formations of four forces. But that's the sun. And how would you be dissecting that lot? Infernal radioactive nemesis be buggered if I'm correlating you here in the sun at its external pleasure. I can feel the lion's technicalities closing around my jugular. Better get out of it while the going's good.

The atomic proton—well the internal dynamics of that, brings us back to the quark rationale and the main point of that whole solar foray. Can the atomic proton spin manifest an asymmetrical build-up, other than through the dynamics of the three other atomic sized force formations? Look, I don't think you can get inside the proton the way you can with the sun. I think the spinning domino effect stops with the atom. So whether your quark can correlate with internal solar activity, I think is questionable. Besides, quarks don't give you the whole low down on spin anyway—let alone any systematic symmetrical estimates. We've probably got more chance of working out how proton spin balances out according to the metaphysical grid of the force and the aforementioned rotational dynamics.

Remember all of that business on calculating force formations according to Einstein's equation, then tacking on Newton's one for the greater gravitational effect? Well, that's what I'm getting at. There could be some figure that symbolises the balance per se. Where your hypo energy and mass equation for spin changes one way—your hypo one for waves could change the other way. Bearing in mind all the time, that spin, spiral, wave or orbital always have a non-negotiable quota of spacetime and motion. Bearing in mind the metaphysical grid. The change in one would affect the equilibrium of another and create an asymmetrical situation. The internal dynamics of spin could amount to a nonlinear equation according to the three other atomic forces and *their* physical positioning going by $E=mc2$. Simple. But who's working all *that* out. Roll on quarks.

On face value, maintaining symmetry means releasing energy whilst controlling a mass increase. Whether we're discretely shoving things into a nucleus and revving up its spin—or more to the point—

simultaneously drawing out its nonlinear fusion and unraveling some of its spin, it doesn't matter. We would also need to regulate the mass situation. And those equations, from the above convolution, would guide us? I think so. We don't want the neutron chain reaction of fission without a simultaneous fusion into motion-mass, for example. Theoretically, we want both. Theoretically, we're probably speeding up one or other sides of a natural chain of events. The energy here comes from manifesting radio-active decay as well as capturing it. And all going well, the stuff coming out should be the more powerful. Our spiralised booster then, is just the prod to unravel full on spherical spinning fusion of spacetime and motion. We might need to bring in extra force formations to do it. We might need neutron assistance. We might need a new brain. I might need the palatable reassurance of a successful brew.

You try delving into the singular spherical spin of say, a proton. Try setting up an aggregate situation *within it,* without numbers or particle bombardments. We've already done the sun business. My mind's warped from the effort. Whoever thinks cosmological theory is just pie in the sky grandiloquence isn't thinking as far as I can tell. Look, this aggregate proton business might not be the same thing as gluing a neutron to it, or some other sub-sub-atomic particle. Could it be so simple? And you can bet your bottom dollar, with today's bent you'd probably end up bombarding it with something. The idea though, is that you'd never end up with a heavier element. The idea is to avoid unnecessary aggregation of the primal spherical situation. So why not stick with the most fundamental and prolific system within the universe—hydrogen? Think of it as the transformation chamber of mass into superluminal energy. How then do you tamper with the hydrogen's proton without transforming it into helium? We'd have to initially recognise those hallmarks of standard science I suppose, namely; the conservation of charge, energy and other constants of infinite and eternal spacetime and motion. Not to mention all the theoretical convolutions.

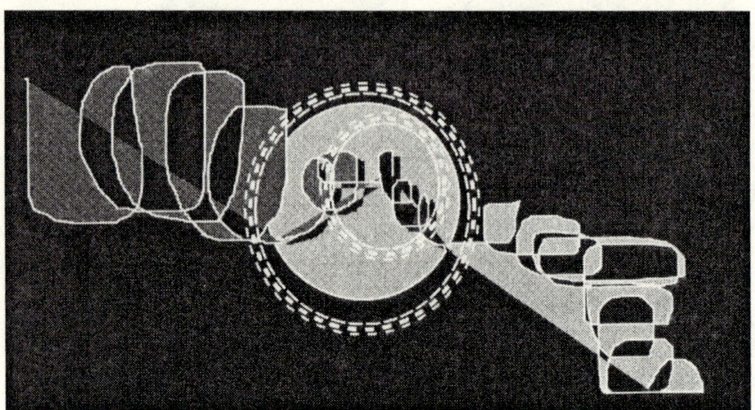

The green spirals in that diagram represent the booster stuff. It transforms with the spherical spinning combination of spacetime and motion, thus building up an excess of strong nuclear force, rendering the formation assymetrical. This adjusts by releasing excess spin according to another spiral—namely the light blue and pink one. This one should comprise of superluminal propulsion energy. So where are the waves and orbitals? Well, I don't know. For simplicity I left them out.

Theoretically however, the entire operation concerns the manipulation of four combinations of spacetime and motion. We're rigging the system. And because we're splitting its processes, or at least honing in on different sides to the one story, maybe we should be addressing two sets of force formations. Over-charging a spin and then releasing its transformed energy, could pit wave-spiral against spiral-wave. We're fracturing the symmetry of the system per se. And we're according a symmetrical estimate to each fragment. I mean; each process should bear some symmetrical weight which then affects the composite symmetry through their interplay. If your wave-spiral is way off kilter the orbital mightn't act the part. And here's the crux of the matter, there's only a single spherical spinning force formation. That's why hydrogen. A lone spherical spin must relate to both sets for symmetrical push and shove. It could be the key to the entire process. The basic symmetry of spherical spin is supposedly always subject to the systematic set-up of spiral, wave and orbital. We're theoretically pumping it up *and* spinning it out, all at once.

In summary then, charging up pure and simple spherical spin involves the waves of electromagnetic activity, gravitational corridors of orbital electrons as well as their spiral passageway of entry into spin. Radio-activating it involves all of that as well. But what ends up coming out of this set-up supposedly surpasses radioactivity as we know it, by a long shot. What comes out should be full on strong force formation power, or the c2 of an E=mc2 transformation. Great isn't it? Here are some theoretical fundamentals, shove them all in there and abracadabra out comes the white lightning. Here we go with the quasar power of a jolly hydrogen to Eridanus and beyond. Have I missed something? Do we really want to go over and grapple with all the ins and outs at this stage of the game? Do you feel like the quark rationale according to the solar correlative, bearing in mind the metaphysical grid? Is it happy hour yet? Nope.

Quantum physics and its zoo, apparently show how emission and absorption isn't continuous but in bundles. The energy levels of all atoms are quantised and only occur at certain levels. Furthermore, quantum particles mediate a force. Well, from what I can grasp of that, all's well and good. Broaden your mind and that can also describe the system of a force formation set-up.

Put the quantum zoo rationale on top of theory and you could be looking at the energy and mass calculations of our metaphysical grid. That might also describe various degrees of nonlinearity thus showing how stuff sits symmetrically. Particle physics could tell us where the energy flow begins and where the mass stops. Put that with theory and it might show up symmetrically and systematically. Look, we've theoretically got four discrete combinations of spacetime and motion at play in any system. The wave will distance the orbitals, etc. And we're trying to manipulate these formations. Now enter particle physics. If we draw the wave into a spiral, for example, through electron removal, we might get a less discrete emission. That's more energy and less mass. And with a continual stream of strong force quanta coming out of its central spin, we want the particles to dissipate into a seemingly continual flow of energy. So which particles do that, and can we come by them via the wave-spiral reasoning?

With all respects Plank, theory seems to present a more holistic picture of force quanta. Look quark quanta increase in mass as the force becomes stronger and more short range. Your long range photons, for example, have zero rest mass. On the other hand, those weak force leptons have some mass and can only spiral around shorter distances it seems, whilst the strong force hadrons might have the most mass and the shortest range. Theoretically, the spherical spin is confining them. And finally, those evasive gravitons probably have the least amount of mass and lose themselves in the greater space and time of a seemingly endless linear passage. No wonder no-one's found them. But the point is, all these force conveying quanta tie in with the metaphysics of the force formation. Greater spacetime and less motion, as in gravity, accords with your weightless ephemeral graviton. Maximum motion and minimal spacetime of strong nuclear force ties in with the relatively massive quark group.

Theory can cover the gaps that the quark reasoning requires. It can link the quark via the gap to the system. Take the immense distance between electron orbitals and the proton. Now that's also the spacetime and motion impact of forces upon each other. This could be the stuff that quantum physics doesn't quantify. Chew up the space and time and we have more nonlinear mass and motion. Take away the electrons and the orbital space corridors don't register. Spacetime and motion could spiral and thus empower our proton. Its spherical spin could then become asymmetrical and release excess fuel. There's a whole range of possibilities here. Shove a neutron in there. With its binding energy and transformative properties, the impact might not transpire into such simplistic energy release. It could absorb systematic spacetime to render proton spin into a compounded mass-motion symmetry. And this could belie any radio-active release of nonlinear power. Either way, we're now looking at it from the added dimensions of force formations. The mass and energy of a quantum particle also relate to the spacetime and motion of atomic symmetry. Each atomic force comprises of a unique combination of spacetime and motion. So we have four different dimensions. All up they lend themselves to energy and mass transformations. All up, they provide a nonlinear perspective. We're getting the symmetrical factor. And this is what could illuminate other interactions in the aggregate atomic set-up. Here is our holistic picture through which to rationalise proton manipulation.

So what I'm saying is that categorising zoo particles according to the spacetime quota of the force *formation* might empower our application of them. Or it might totally confuse you. That is; if due to some window of inadvertent lucidity, it hasn't already. Quantum particles increase according to the degree of nonlinearity of the force. You get more with the strong force then out of a wave, for instance. I'm hoping we might see why some transpire so readily into other ones. I'm hoping we might see it for the sake of force formation manipulation.

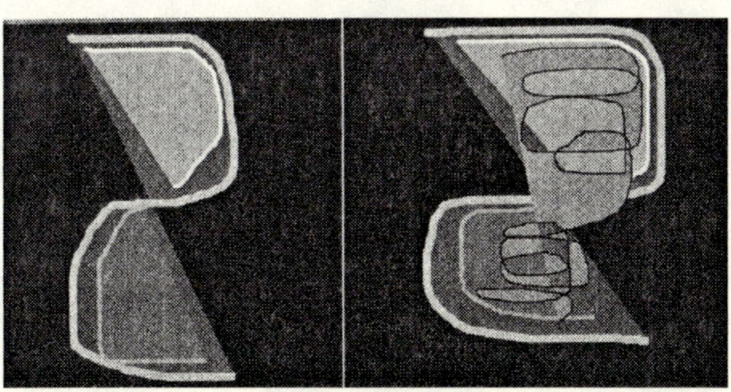

Here are examples of quantum zoo particles within an electromagnetic wave. Superimposing the spirals on the wave show the increases. The top half of both waves comprise of their 'positive' particles with pink anti counterparts below. The blue spirals on the right show an increase in particle quantification within the speed c limit of the wave format. These would be our weak force leptons. You can see the difference in the space content of the wave particles compared to them. The latter would have more nonlinear mass-motion and less spacetime.

Anyway the more mass, ostensibly the more short-lived is the force particle. So enter theory. If the particle's mass is subject to a force formation it would also represent that force's degree of nonlinearity, or a fusion of mass with motion. Indeed, should we substitute mass for motion we'd probably come up with the appropriate combinations of spacetime and motion relevant to each force formation. Your massive hadrons of strong force spherical spin would have minimal spacetime and act accordingly. Zappo. We want a mass-motion release into energy out of the strong force formation, don't we? And the hadrons look to be the likeliest candidates. The hadron transience of mass could transpire into a flow of energy. But could they spiral out the heart of spinning sphere power accordingly whilst some other little number is winding it up?

What's more, your particle transformations could reflect formation transitions. I don't think force formation particles are as 'cut and dried' as the zoo would have them appear anyway. The weak force leptons ostensibly do more than spiral. You've got your electron relevance which brings in affiliations with orbital corridors. In fact all of these zoo particles could have other force formation affiliations. It's one for all and all for one in the atom anyway. How and what they transform into, that's the system's symmetry stepping in. Or do they just reflect the tangents and nuances of a particular force? Let's drag out the neutron again. What a classic case of chameleon talent that is. There's your whole atomic system potentially within. Whereas, some zoo particles, with their propensities for transforming into flavoured relatives might just be showing up quantum slices of spin. We're talking the recoil evidence of the force's shock tactics with them. Your neutron, however, is a natural event. All particles might bear some abilities to do this and our neutron in its natural wonder just exemplifies the kit and caboodle of such brilliance. The flotsam and jetsam we're grappling with could still carry traces of it. That's not the point, however. The point is when they change. The charm of one and the colour of another might be a bent towards a spiral come wave. Put that with the particle transformation—what the jolly little beasty transpires into, and we might discern another formation affiliation.

There's loads of them, you know. And thus far they're probably only defined as bits and pieces of positive and negative quantum according to speed c. If we're to use the quark rationale effectively here they could *all* require theoretical translation. Be buggered if I'm doing that though. I wouldn't know the difference between a pion and a muon and a Zeppo. The idea, however, is to manifest a set of them by way of force formation activity. And do we need the whole set anyway? If we appropriate some bits and pieces of force formations according to the system's zoo logic, that might be enough to get our system sufficiently off kilter for the desired result. The more simple—the more possible. But to grasp the simplicity—to know which piece of quark would up-end the system, shouldn't we first piece them together?

And if you're thinking what a baffling puzzle already this hypo superluminal set-up is, then just imagine cutting to the chase in a composite nucleon. What a tangle. The closer we are to simple spherical spin, the more access we have to that short-range strong stuff of c2. By tunnelling into the pure proton with a quantum sibling, we might set-up a spiral access system for hadron emission. Even from the cathedral roof of atomic space we should be able to skirt around the orbitals and wave-spiral in tandem with spiral-wave accordingly. So move over Cirque du Soleil. I'm coming in.

Too bad I'm scratching my head and thinking muons, puons—they're not for me. Who knows, some genius understanding these things might crack their code with theory. That's what it's all about. How can you be up above the forest and figure it through the technicalities of its mossy undergrowth? Immersed in the depths of involuted reality, could theory travel this far? Besides, my vantage point only allows reasoning the trees from the irreverent perspective. I'm just putting it to you that what I'm getting out of it should also add up technically. What those details are, however, are clearly beyond me. I'm not *that* irreverent.

What then have we got? A hypothetical process of proton power. And what's believable about it? By starting with the hydrogen proton, and working it according to four force formations, I'm addressing our most basic, and prolific, symmetrical system in the cosmos. At that level you could get the symmetrical edge. A basic hydrogen system may not stand for any further internal systemisation. Quarks, skuarks or otherwise, possibly only add up to this bottom rung of the bifurcating ladder. I mean I don't think you'd ever get a full force set-up inside a quark. And even if you did I'm judging nothing by them. I'm sticking with proton lucidity and what the quark means according to that. It's that and the neutron.

If we approach it with a neutron would the proton transpire into another 'hydrogen' set-up? The 'free' neutron decays into a proton, electron. . . . Would the neutron just compound the system, and/or reproduce it? And I don't want to transform hydrogen into something else. With respect to the conservation of everything I just want to use the nuclear *mechanics* of our most basic set-up. Put it in, spin it up and churn it out. And let's get out of our bubble.

Can we do that with the nucleons of neutrons? Could we rev up a spherical spin through nucleon activity? Would they spit out transforming strong force spin? Would neutron-proton leverage influence spin velocity and spiral formation? Maybe it's like removing the casement off the pivotal driving force. But could we still get the simple emission of excess spin this way? Or would it just transpire into a fission chain reaction or elemental transformation? The thing is with neutrons you're covering both sides of the spinning coin. They could have the bilateral flexibility for both radio-activity and its nemesis. And look your nucleons can absorb energy from outside sources and enter excited energy states. Indeed, their rotation and vibration apparently account for their excitement. We'd still have to look to the symmetry of the system though.

But I bet after all of this—I bet you don't believe all of this symmetry and systems business. I bet you think the bifurcating ladder is bullshit. Let me take it from the top then. All atoms decay don't they—at some rate and stage. A given number of them should decay in a given time proportional to the total number present. So even if we don't tamper with the hydrogen it will eventually overcharge and unravel anyway. And who's to say that's not the overarching grand-scale symmetry of cyclical events? There's your stellar systems and their symmetry, and galaxies happening together with the expansion and contraction rates of the cycle. I'm just calling the shots on a natural event by order of systematic symmetry. I'm gauging the atomic system's forces to fire away all at once in respect of the greater deal. And I think it all comes down to three main points. Firstly, we use the mechanics of spherical spin to generate power. Secondly, we appropriate spin mechanics through the symmetry of an atomic system. Thirdly, keep it simple.

So there you have it. I've reached my suggestive limit. One thing, however, remains sure. Despite all the loop holes, discrepancies, and technical ignorance, theory here transcends fission, fusion and any other proven nuclear process as we know them—from the word go. It presumes speed c squared from the on-set. It's already there in the proton spin. Theory just hones in on this core of power, holistically. On that basis, it's poised to carry these procedures, and any other ones you can figure out, along the superluminal pathway. We just need to confirm it's existence and then tap into it. And now it's enough with the theory. Anymore and we're past the point. We'd be well and truly in the realm of science fiction. And who's going down that road with this. No, thank you. They're still blowing each other up in the movies. And may that force remain with them, in celluloid land.

. endnotes

1 nb translate that as minimal length and breadth of Newton's acceleration

2 nb amazing how the seasons muddle up when you edit in and out stuff with each

3 nb I know that the strong force has minimal reach in terms of spacetime per se, but that's not to say its force formation isn't subject to the symmetry of the system's 3 other ones. Indeed that's the point. These four formations are a symmetrical division of their total.

ch.10

conclusion

. are we ready for it?

Just working out the mechanics of spherical spin probably wont
make it happen, you know. Even if, and heaven help us, there is
something in it, it's not enough. And should Speedy Gonzales
descend from the clouds and hand over the low down on inter-
galactic travel, it probably wouldn't register. I mean there could be
morally more to building a space-ship according to superluminal
specifications for blasting off into galactic society.

For starters, there's probably a host of philosophical issues between
us and the greater cosmos. Philosophy apparently affects the human condition, particularly the attitude. And I'm wondering that
the present ones are either useless relics flapping through the winds
of yester-year, or the gale force blasts of contemporary cynicism.
All of which could dispose us towards extinction rather than cosmic discovery. I reckon we're in the wrong frame of mind. Where's
the general mettle for getting out there? We've still got the aggressive nature. It's alive and kicking and not just creeping through
the annals of human history. We still don't have the sagacity to
overcome our own wars. And there's a conscious choice in this.
Shall we blame the human condition for it? I hope so, because
that's subject to change, and increasingly changes according to
our own doing. What I'm getting at is the power of philosophy
with politics. To date I can't think of any that's stimulating cosmic
curiosity.

Philosophies seem to guide and reflect our evolution. Look, I'm no historian, but I don't know of any meaningful time in human history without some philosophical input. They seem to link in with our realities. And without one covering the greater cosmos, our realities could sink in earthly aggros and a spacetime compression of modernity. That's science and technology too you know. I mean philosophy seems to give scientific observation and experiment the room to move and cogitate. The philosophy of a greater world can at least inspire you to an awareness of broader horizons. If you're not aware of broader horizons and reaching out there, where's your impetus to rise above the immediate in your techno drive? If we're not reaching for a rational position in the beyond and realising things according to it, we could get swamped in our own realities. As sure as tomorrow, they're going to change with or without you—or me, or the white wallaby, or anyone you know. And so says Laurelle, the font of all effrontery. Heavens, even Democritis, in figuring out the atom, was trying to construct a philosophy of the material universe as well as of man and society.

Perhaps at any given age, we understand so much space within a certain amount of time, through so much motion as reflected through our realities of the day. And all of which could be for some reason of our own devising. Take the spacetime and motion of human experience today. All up, we're probably as great as the nocturnal orbital of a satellite. We're as great as the speed of gravity. It defines our cosmic play ground. Theoretically however, the possibility for superluminal travel derives through the eternal and infinite nature of the cosmos. On that basis, if you do not *accept* the universe is infinite and eternal you probably would not arrive at a rationale position for getting very far. You wouldn't have the dimensions for calculating it. Nor would you have the philosophical inspiration to do so. Clearly, we can't observe an infinite and eternal cosmos. So it could take a philosophy of one to make initial sense of it. Indeed, it looks like obtaining superluminal velocities is also a means for grasping the greater macrocosm—for working our the infinite and eternal nature of the cosmos, should it so be.

But just try constructing one such philosophy. We're supposedly covering everything that you can, conceivably and non-mathematically. There's; no beginning, no ending and it goes on forever. Who's believing that, let alone some unqualified theory of it? And with all due respect to theology, belief doesn't come easy these days. Remember the real world ethic. Seeing is believing. Now if you can't at least believe or imagine an eternal and infinite cosmos, how then are you going to form a philosophy for it? You might as well have a hole in the bucket with Henry trying to fix it, dear Liza. . . .

Clearly the entire universe extends beyond Hubble's telescopic range. So most of the stuff is inaccessible to us. So you believe it's there. But we don't, you know. All I reckon we're doing is judging its existence according to an extrapolation of the observable. Belief, schmelief. We're rapidly replacing belief with knowledge. And that's knowledge according to the power of us and our human eye sight. On that basis, any extrapolation just extends the expansion of the below speed c range. You don't get out of Hubble's bubble with it. Once upon a time though, in the land of ancient philosophers, belief was part of the knowledge process. Along with imagining and reasoning, believing formed the function of understanding. Well, it did for Plato. And I suppose it did for democracies. But today, well what's guiding them now? Can the present systems lead anyone to deduce far reaching principles and greater objectives let alone inter-galactic travel?

You might as well go and play with the pixies at the bottom of the garden than contemplate cosmology according to all its potential mysteries. We've possibly got a super-abundance of the below speed c scientific method on the road leading to anywhere. And what's not up for such scientific verification probably doesn't register. Well, it wouldn't, would it? Anyway, I can't see worm holes, time tunnels, and the current black hole business broadening anyone's outlook. I don't see them developing any values and guiding society through a depth of vision and scope towards greater evolution. Indeed cosmology, as an addendum to our limited observation and experiment, is probably a waste of time and money to most people. It could mean as much as a forty day fast in the desert does to the standard anorexic. I mean get real; time's for me—now. It's not eternal. Time is money and it pays to advertise. That seems to be the current philosophy.

So it's probably pointless looking to philosophy to guide you up there. They say there's an impasse between science and philosophy, anyway n b1. Those twin pillars of human reasoning are apparently no longer complementary. And I reckon while science is bounding ahead up to speed c, philosophy is the sad one spluttering in its wake. A scientific takeover, you think? Perhaps postmodernism, realism, capitalism—and let's face it—even orgasmic hedonism, are just derivatives of the scientific method translating through *isms* into philosophy. I don't see any of them forging avenues into the greater unknown for thinking. Indeed, cosmology could primarily be the number crunching cold-face of science.

Where then is your means for developing objective awareness? Mine seems to be going out the window. Just writing about this stuff is gloomy. And it's jolly Christmas today as well, you know. I'm waiting for the turkey to cook so we can load up the haversacks and head into the forest with all the goodies and a gold pan. There's that dell down there near that trickling creek that's always sparkling with mica, and it's got all those big tree ferns, myrtles, sassafras and mossy spaces surrounding it. Hey ho, for Christmas with the wallabies and the home brew then. But for now I've talked myself into the pessimism of the subjective spacetime compression. Well, the turkey needs basting and the fire's going out, so here comes some immediate escape. Hang on.

Where were we? Oh yes, forget accepting any eternal and infinite possibilities of a greater cosmos, that's light years beyond the rivulet of today's objective rationale. Sure, 'objectives' figure prominently in any boardroom. I mean it's the pro-active fashion of public service departments and boardroom meetings to outline them. And well and good. And so it goes. But where does that leave cosmology? I don't see that rationalising free market forces of a global economy is on the road to the stuff I'm on about. Or that pumping iron to a target realises any greater good. Subjective values are more than likely driving any objective thinking, if it's coming from the proximity of realism. Look our realism tends to concern the here and now more so than any position way over there. Where's our framework for the longer term perspective? I know we've done paradigms, but where's one applicable to an eternal and infinite spacetime and motion? Are we forgoing an ability to shape reality by mis-representing its capacity to determine us? Take nuclear deterrence. It looks less ethically objective, and more ominously subjective, every time a new member joins the club. Power to the subscriber.

In any event, with no philosophies framing the deep future of cosmic speculation, you'd have to be more subjective in your thinking. You could end up as a big fish in a little pond with scarce opportunities for anything other than mutated deformation. And how would your descendants be? Cloned. They mightn't give a hoot, well, less of one then what's going around now. Which is looking very dodgy if you're contemplating unfurling superluminal power. And where's your global perspective coming from? No wonder we still haven't figured out the planet. Small wonder we're still working out the weather and the fish. Bugger arriving at the cyclical rationale. For us, the variety of factors involved don't tally up with any order in nature. What order in nature? Where's the context for working it out? We're coming at things from an egocentric viewpoint inside a cosmic bubble. Your cyclical order of nature—

theoretically, takes more time, space, and motion than we're recognising.

. *setting the stage*

It could take billions of years for life like ours to evolve. That's a lot of time, *a lot of space* and a lot of motion going on. Enough that is, for all the symmetrically, interacting and bifurcating factors to sprout brain power. You'd have to have a star and some planets happening in a stellar system according to the symmetry of a galaxy formation, it seems. And it would only work out subject to your general relativity ruling the roost and stabilising stuff down. Theoretically it's all a matter of symmetrical inter-connections and transformations afforded by an eternal and infinite agenda. We're conceivably what comes off the conveyor belt through the bifurcation of four forces of spacetime and motion. When the conditions add up, out we sprout. I mean stars translate into people, don't they? People probably translate into stars. And you need your star to stay alive n b2. Well, we know they come and go. You'd reckon the jolly star, if nothing else, would add up for us according to galaxies and encourage the cyclical rationale. But it doesn't. We don't apparently come and go. We're the only all time cosmic civilization, going by us. And we're stuck in the all time big bang scenario according to Hubble's bubble and COBE's evidence. The cosmos came out of a pin head that proceeded nothing.

Look at the solar system. This one could be a special cosmic cranny ripe for human existence. How do we know its other planets didn't house earlier life? How do we know that Martian microbes are all there is to that? Your Martian microbe might've marched down the same road into nuclear technology. And whether we came out of a fish, a spaceship, a monkey, or crawled indirectly out of the earth's core, we're still subject to the solar system's elements. Look hard enough at the other planets in our solar system, and they start resembling the remains of planetary holocausts. Surrounding us could be the burnt out shells of previous existence. They could've sustained other subjective creatures—blind to so many avenues—down some inevitable road that we're on. Maybe it's in our genes. Maybe your supernova—come black hole—come general level cyclical junction even, doesn't sufficiently re-arrange the potent atavism of human aggression to prevent its re-occurrence when the conditions are right. Maybe no amount of spacetime and motion is great enough for that. Hypothetical or not however, the fact remains; no-one seems to be stuffing up the planet except us. We can blow it up at least 27 times over. Great, isn't it. Talk about the big bang.

Thus far our life in this solar system seems more finite than its solar spin. And gathering the knowledge to transcend it, other than through re-arranged star dust, may depend on our values and how they formulate. Without a different way of thinking the quest seems useless. You know where I'm coming from. What I'm saying is that objective knowledge is probably the way to go. And it probably comes from the bigger paradigm with the scientific-philosophical room to move. Anybody knows that the objective rationale is a powerful instrument for achieving certain ends. Just take it further than we generally do. You could find a cyclical pattern to life in reasoning things through the bigger picture. You can ask different questions then. Why are we in cosmic isolation below the speed of light? You can question the speed c limit and other barriers that prevent cosmic exploration. You can draw a line in the sand and question what's over there. Then you can step over the boundaries of observation and experiment and look back at what's over here—through a different light.

We could undergo some sort of 'planetary self-transcendence' (Lance Morrow's term) by reaching into the galaxies over the speed c barrier. It might put us on the morally correct avenue—the one branching off planetary burn out. Even so, the most utopian idealist couldn't possibly conceive of life and everything just gearing up towards getting around superluminally within infinity and eternity. How would you be, forever on the go, up there in the stars and beyond? Still the point remains, getting up there could be helping us down here.

And now we're at the end of the book I might as well be academic about it. I've got something here gathering dust from my university days. It's about developing objectivity. I'm talking about our political structures and our war mentality. Without addressing political realities nothing much happens—cosmologically or otherwise. So I might as well shove it in. Look, philosophy in any context seems largely ineffectual if it doesn't translate through political realities. Your political systems enact your ideology. And all things being relative, they should at some stage influence cosmology, although I would suggest that it's a two way thing, with the political side of the planetary coin evidently the more powerful. Anyway, what I'm getting at here is how our political systems can also influence our way of thinking.

At university, the concept of 'multilateral fora' arose as a point of interest. It's a way for ideas and reality to translate through each other in the broader framework. In retrospect it's probably akin to a U.N. devolution alongside its centralization, if there's any such animal. And in various forms it's happening in isolated pockets already. You know—your government gatherings, commonwealth meetings, South Pacific forums—all that. And several steps beyond them, a genesis of 'multilateral fora' connected the philosophy and political science faculties for me. It even walked around the Middle Eastern one and entered into Qu'ranic translations. I took it as a way of stopping wars and advancing human evolution. I thought if you connected political systems through a multilateral set-up, you'd get the objective framework right up front. Your state problems would then align with your global concerns. And your state value could form through the bigger picture.

I was thinking that a scientific and philosophical agenda should have more say. And that a multilateral political arena would allow for this. The reasoning being that 'multilateral' doesn't only mean the geographical terrain. It also refers to the rationale of issues. Your issue goes within the comprehensive context. You get the long term rationale in a bigger dimension of comparative side issues. In that context nuclear technology, for instance, might mean more than Nato membership. It could register cosmologically. But as things stand today, cosmological research is still up against state warfare. And doubtless, space funding loses out to smarter bomb research. Our rivalry surely affects any objectives, scientific or otherwise. If on the other hand, everyone's answerable within a multilateral set-up, the rivalry might not be so prevalent.

Unfortunately, you've got your historical trend of states struggling for power. Habits die hard and trends remain difficult to overcome. It's still happening with our present systems. Just ask a Serb in Zahgreb, an East Timorese, a Kashmiri, or a Chechnyan. Look at the British when they rallied around their flag as they geared up for the Falklands take back. Without your common enemy, nobody is uniting. And sometimes it seems that without a struggle there's no point. Either way, domination and aggrandisement still define many state goals. All of which comes about via conflict. Enemies, and the need for them is a motivational force. And I don't see that ringing the creed of any prophet would overcome that. You're probably wasting your breath expounding the virtues of monotheism today. Unite in the face of one God—what God? The God that's eternal and infinite. Clearly these age old power struggles are more to the point. We're still using political structures that require an enemy to justify existence. Statehood and political independence generally means overcoming your neighbour.

The thing is, while all of this is going on, your global concern and objective rationale is whistling in the wind, if indeed it was ever

there in the first place. Never mind cosmic awareness. In the meantime, international politics progresses through the subjective form
of reasoning. It all translates into reality. States go on rationalising,
and are rationalised, through weapons systems and worst case
scenarios.

In this manner policies keep on creating conflict. You end up with
a divided world capable of self destruction, which is what we've
got. States require arms proliferation. Government policies then
allocate funding for research and development accordingly. And so
it goes. The ball keeps rolling and encourages a particular form of
scientific thinking. It's not the objective form I've put up there on
the multilateral pedestal. You're looking at the scientific method
devoid of any cosmic philosophy. And the realities unfurl. Development comes to rely on technologies that realise destructive forces.
You get it in every avenue of society. *It's a dynamic force that's difficult to dissipate.*

Plus we're in the 'me-now-very-fast' era. Maybe your long term
values and consequential reasoning only apply to the size of a chess
board, or a sporting target of bonus muscle. The subjective gratification a delusion for the objective rationale. We've got things happening on the planet now that no one is in any position to address
or resolve. And if, by some chance, we get it right on something
where's the political will to do it? Despite Nato and the U.N.,
individual state concerns can still overwhelm the general good.

One way out, that I can tell, is to put the subjective value within
a context of multiplicity, where power devolves through a multilateral set-up. You get more power that way. I even went into the
metaphysics of it then—the causal phenomenon of power. The
more interactions something sustains the more power it manifests.
And maybe that's your spherical spinning situation in a sense,
however . . . The state multilaterally gets more choices. Different
values start happening and power disperses comprehensively. It
means when the bubble of immediate concern enters a deeper

dimension the possibilities for objective thinking are greater. So the state head, for instance, can then handle a decision through an awareness of the state within the world, that being a new level of understanding.

Ostensibly, it's a way for knowledge to develop creative forces rather than destructive ones. I think it's a way of stopping wars. And that's multilateral fora for you. It could do a lot for cosmology. Getting the global issue on the state agenda is after all, on the road to the bigger picture. A global set-up like that could elucidate an order of nature hitherto unknown, for our comprehension and future evolution.

How else can you predispose the objective rationale politically? Without full participation in a body like this, how would you get effective international legislation on things? You might see traces of it happening around the place. But without the combined political will, state fish remain buggered, state forests keep burning, everyone's smoke keeps rising, wars go on, not to mention the starving billions, the over-population and the struggle for survival. I'm facing reality here. And with this multilateral addendum I'm trying to show that the philosophy behind the cosmic concept recognises hard core reality—albeit at some blatant level. Here is a structural set-up with a causal effect. You get a solid political structure that can direct subjective values into the objective arena. All up and eventually with it, you could be on the way to cosmic awareness.

It's no secret that we can develop different values anytime we want. And given time and structural guidance certain values should become visceral. If so, we've got the potential to go by instinct onto a higher level or egress according to the aforementioned atavistic aggro ones. The idea is to develop awareness. A cosmic one might lift humanity beyond its rising hedonism, cynicism and subjectivity. How many more declarations of independence do we need? How free have you got to be to know where it's at? It's a big thing.

Political and philosophical changes don't just happen. But then, nor do perceptible superluminal velocities.

Doesn't one's attitude to society and the cosmos precede any major scientific breakthrough? Shouldn't we value our world and see our place within it as a precious gift rather than a given right? What do you want; cities of cars or oxygen? Perhaps we've created a situation that now requires a choice. And if we don't choose maybe nature will. We could be star dust sooner rather than later by disregarding nature. On the other hand, nature might fly us around the universe in the twinkle of an eye, should we come to understand it.

. endnotes

1 nb they said that when I was at the ANU. (Australian National University)

2 nb even if some other civilization dumped us here as convicted miscreants, the spacetime and motion factors should still apply. That is; the symmetrical positioning of spacetime and motion necessary for us to exist, could mean that you'd never get a suburban sprawl of humanity across the cosmos.

definitions

a priori: known to be true independently of experience of the subject matter relating to or involving deductive reasoning from a
general principle to the expected facts or effects

bifurcating ladder: the symmetrical scale of force formations ranging from the atomic level to the cyclical one of general relativity

calibrate: to determine the accuracy of something

concept: general idea

cosmic: of or relating to the whole universe which is, in this case,
eternally and infinitely extended

cosmology: the science concerned with discerning the structure
and composition of the universe as a whole using astrophysics,
philosophy, etc., and in this case theory and a belief in eternity and infinity

cosmos: the universe considered as an ordered system

Cygnus X-1: an intense source of x-rays that lies in the direction of
the constellation Cygnus, and almost certainly contains a black
hole. The x-ray source is in orbit around the star HDE 226868
in a binary system, and has a mass of 6 to 15 times the mass of
our sun, well above the Oppenheimer-Volkoff limit

dark matter: hypothetical matter that, according to current theories of cosmology, makes up 90-99% of the mass of the universe but so far remains undetected

delta pavoris: 19 light years away

determinism: the philosophical doctrine that all acts, choices and
events are the inevitable consequence ot antecedent sufficient
causes

eternal: a name applied to God's infinite wisdom
without beginning or end

extrapolate: to estimate beyond the known values by the extension
of a curve

flywheel: a heavy wheel that stores kinetic energy and smoothes
the operation of a reciprocating engine by maintaining a constant speed of rotation over the whole cycle

gravitation: the force of attraction that bodies exert on each other as a result of their mass

God: without beginning or ending

helium: an inert gas

incandescence: emission of light from a substance in consequence of its high temperature

infinite: having no boundaries or limits in time, space, extent or magnitude

krypton 86: the calibration of most scientific insruments is through the measurement of length based on the wavelength of kr.86

light: the sensation experienced when electromagnetic radiation within the visible spectrum falls on the retina of the eye

luminescence: emission of light from a body when atoms are excited by a means other than raising its temperature

magnitude: the brightness of a star or planet expressed on a scale in which lower numbers mean greater brightness
a linear form of measurement

metaphysical grid: the unchanging combination of spacetime and motion of each force formation

metaphysics: that branch of philosophy that deals with first principles, especially of being and knowing

nonlinear: having more then one dimension, circular and spherical

paradigm: a working pattern, model or representation of a concept or theory

Pauli exclusion principle: 2 similar particles cannot occupy the same quantum state

philosophy: the rational investigation of being, knowledge and right conduct

photons: a quantum of electromagnetic radiation with energy equal to the product of the frequency of the radiation and the Planck constant

predicate: to base a proposition
to assert as a condition of the subject of the proposition

physics: the scientific study of the interactions of matter and energy based on mathematics

Planck constant: a quantum theory accounting for the spectrum of a perfect radiator

a fundamental constant equal to the energy of any photons divided by its frequency

postulate: assume to be true

prerequisite used as an initial premiss in a process of reasoning

quality: a distinguishing characteristic

the basic nature of something

quantify: to discover or express the quantity of something

quantize: to restrict to one set of values

quantum: the smallest quantity of some physical property that a system can possess acording to the quantum theory

in order to explain the phenomena of emission and absorption of radiation Planck said that it was emitted and absorbed in quanta a particle with such a unit of energy

quasars: quasi-stellar radio sources that are very distant objects like stars, but are separate island galaxies radiating amounts of energy by processes not yet explained by any known laws of physics probably several thousand million light years away

radiometry: the measurement of transfer of energy of radiation by light

radiation: the emission or transfer of radiant energy as particles, electromagnetic waves, sound, etc.

rubidium: has a half life of 47 billion years

scalar: having magnitude but not direction

sooty: lovable very, very, verry cheeky-naughty black cat with beautiful eyes; chirrups, purrs, climbs trees, wanders miles, eats moths, mice and dainty tiddies, sleeps, rolls in the dirt, hides in the grass, plays with the wallabies

space: the unlimited 3 dimensional expanse in which all material objects are located

spatial: an interval of distance or time between 2 points, objects or events

spectrum: a record of the distribution of matter or energy (light) by wavelength

spiral: one of several plane curves formed by a point *winding about* a fixed point at an ever increasing distance from it

a curve that lies on a cylinder at a constant angle to the line segments making up the surface

spiral galaxy: galaxy consisting of an ellipsoid nucleus of old stars from opposite sides of which arms, containing younger stars spiral outwards around the nucleus

star: rises on average 2 hours earlier each month

superluminal: faster than light

symmetry: balance among systems or parts of a system

a state or system that has a significant quantity that remains invariant after a transformation

tachyons: various superstring theories allow the existence of particles that can travel at superluminal speeds. Tachyon is the generic name given to them. Tachyons perform time travel, which permits travelling back in time. Most physicists don't support thoeries that involve tachyons, and there's no evidence of them.

theory: a plan formulated in the mind only

a system of rules, procedures, and assumptions used to produce a result

a set of hypotheses related by logical or mathematical arguments to explain a wide variety of connected phenomena in general terms

time: the continuous passage of existence in which events pass from a state of potentially in the future, through the present to a state of finality in the past

a quantity measuring duration, usually with referance to a periodic process

time is considered as a 4th co-ordinate to specify an event

transformation: generally—an alteration in form, function, energy, etc.

mathematically—a change in the position and direction of the referance axes in a co-ordinate system without an alteration in their relative angle physically—a change in an atomic nucleus to a different nuclide as the result of the emission of either an alpha or a beta particle, that itself is a helium nucleus emitted during a radioactive transformation

basically—ongoing

bibliography

G.J.Aitchison. *General Physics.* London: Chapman and Hall 1970

A.Angelopoulos. *Will the Atom Unite the World?* London : The Bodley Head 1957

Isaac Asimov. *Atom.* London: Mandarin Paperbacks 1992

J.Baggot. *The Meaning of Quantum Theory.* Oxford Univ.Press

J.Barrow & F.Tipler. *The Anthropic Cosmological Principle.* New York: Oxford Univ.Press 1986

D.Bergamini. *The Universe.* Nederland: Time-Life International 1964

D.Bohm & B.J.Hiley. *The Undivided Universe.* Routledge Paperbacks

H.Breuer. *Physics for Life Science Students.* New Jersey: Prentice-Hall Inc. 1975

ed.S.Cahn. *Classics of Western Philosophy.* 3rd edition 1990

N.Calder. *The Key to the Universe.* London: BBC 1997

ed.J.Carey. *The Faber Book of Science.* London: Faber & Faber

E.Chaisson. *Universe.* New Jersey: Prentice-Hall 1988

R.Clay & B.Dawson. *Cosmic Bullets.* Australia : Allen & Unwin, 1997

R.Clark. *Einstein the Life and Times.* London: Hodder & Stoughton 1973

Ed.P.Coles. *The New Cosmology.* Cambridge: Icon Books 1998

P.Covenay and R.Highfield. *The Arrow of Time.* G.B.: W.H.Allen 1990

CSIRO. *Radio Astronomy.* Canberra: 1985

P.Davies. *The Edge of Infinity.* London: Penguin Books 1981

P.Davies. *Other Worlds.* Penguin Books 1988

Ed.P.Davies. *The New Physics.* Cambridge Univ.Press 1989

P.Davies. *The Mind of God.* London: Simon & Schuster 1992

Paul Davies. *About Time.* England: Orion Productions 1995

P.Davies & P.Adams. *The Big Questions.* Australia: Penguin Books 1996

D.Deutsch. *The Fabric of Reality.* London : Penguin Press. 1997

A.Einstein. *Ideas and Opinins.* New York: Crown Trade Paperbacks 1954

T.Ferris. *Coming of Age in The Milky Way.* G.B.: Vintage Bodley Head Ltd 1989

Ed T.Ferris. *The World Treasury of Physics, Astronomy and Mathematics*. USA : Little Brown & Co. 1989

T.Ferris. *The Whole Shebang*. London: Weidenfeld & Nicolson. 1997

M. Freeman. *The Guide to Space Travel*. London: New Burlington Books 1961

J.Gleik. *Chaos*. London: Cardinal 1987

S.J. Gould. *Life's Grandeur*. London : Jonathon Cape 1996

John Gribbin. *Companion to the Cosmos*. London:The Orion Publishing Group Ltd. 1996

S.Hawking *A Brief History of Time*. G.B.: Bantam Press 1988

S.Hawking & R.Penrose. *The Nature of Space and Time*. New Jersey: Princeton Univ.Press 1997

W.Heisenberg. *Before the Beginning*. England: Penguin Group 1989

T.Hey & P.Walters. *Einstein's Mirror*. Cambridge Univ. Press 1997

C.Hogan . *Primordial Deuterium and the Big Bang. Scientific American* vol.275 no.6. NY: Vantage Press Inc. Dec 1996

J.Horgan. *The End of Science* . G.B.: Little Brown & Co. 1996

M.Kaku. *Hyperspace*. N.Y.: Oxford Univ.Press 1994

WJ Kaufmann 111. *Black Holes and Warped Spacetime.* San Francisco: Freeman & Co

C.Kilmister. *The Nature of the Universe*. London: Thames & Hudson 1971

H.Kueppers. *The Basic Law of Color Theory*. West Germany: Barrons 1982

Ed.P.Lafferty & J.Rowe. *The Hutchison Dictionary of Science*. Helicon Publishers 1993

L.Lederman. *The God Particle*. London: Bantam Press Transworld Publishers 1993

R.Lewis. *Space Exploration*. New York: Lansdowne Press 1983

Ed.S.Mitton.*The Cambridge Encyclodaedia of Astronomy*. London: J.Cape Ltd. 1977

P.Moore. *Atlas of the Universe*. G.B.: George Philip Ltd. 1997

R.Morris. *Achilles in the Quantum Universe*. London: Souvenir Press 1997

G.Murchie. *Music of the Spheres.* London: Rider & Co.1979

I.Nicholson. *The Road to the Stars.* Australia: Cassell 1978

H.Pagels. *The Cosmic Code.* New York: Simon & Schuster, 1982

M.Ridley. *The Origins of Virtue.* Great Britain: Viking, 1996

M.Rees. *Before the Beginning.* GB: Simon & Schuster, 1997

M.Rowan-Robinson. *Cosmic Landscape.* Oxford: Oxford Univ.Press 1979

Sears & Zemansky. *University Physics.* Massachusetts: Addison-Wesley Publishing Co. 3rd Edition Part 2 1964

G.Shortley & D.Williams. *Elements of Physics 5th Ed.* New Jersey: Prentice-Hall 1971

L.Smolin. *The Life of the Cosmos.* London: Weidenfeld & Nicolson 1997

G.Smoot. *Wrinkles in Time.* G.B.: Little Brown & Co. 1993

C.Snow. *The Physicists.* London: MacMillan Press 1993

J.Taylor. *Black Holes: The end of the Universe?* Souvenir Press 1973

M.Velimirovich. *The Atomic Universe.* London: Routledge & Kegan Paul 1974

John Wiley & Sons. *Physics for Students of Science and Engineering* New York : 1960

ed.C.Wilson. *The Book of Time.* Australia: The Jacaranda Press 1980

G.Zukav. *The Dancing Wu Li Masters.* G.B.: Rider & Co 1979

Cosmic Coincidences. Black Swan Breat Britain 1991

The Discovery of Subatomic Particles. N.Y.: Penguin Books Freeman & Co 1990

90000
9 780738 828626